NATURE ALL AROUND US

NATURE ALL AROUND US

A Guide to Urban Ecology

Edited by

Beatrix Beisner,
Christian Messier, and
Luc-Alain Giraldeau

Translated by

Beatrix Beisner

THE UNIVERSITY OF CHICAGO PRESS
CHICAGO AND LONDON

Beatrix Beisner, **Christian Messier**, and **Luc-Alain Giraldeau** are all professors in the Department of Biological Sciences at the University of Quebec at Montreal.

The University of Chicago Press, Chicago 60637
The University of Chicago Press, Ltd., London

Illustrations by Alexandra Westrich.
 Published 2013.
Printed in the United States of America

22 21 20 19 18 17 16 15 14 13   1 2 3 4 5

ISBN-13: 978-0-226-92275-1 (paper)
ISBN-13: 978-0-226-92276-8 (e-book)

Originally published as *L'écologie en ville*

Library of Congress Cataloging-in-Publication Data

L'écologie en ville. English.
Nature all around us : a guide to urban ecology / edited by Beatrix Beisner, Christian Messier, and Luc-Alain Giraldeau ; translated by Beatrix Beisner.
pages : illustrations ; cm
ISBN 978-0-226-92275-1 (paperback : alkaline paper) — ISBN 978-0-226-92276-8 (e-book) —
1. Urban ecology (Sociology) I. Beisner, Beatrix E. II. Messier, Christian C. III. Giraldeau, Luc-Alain, 1955- IV. Title.
HT241.E36413 2012
304.2'7—dc23
2012005527

♾ This paper meets the requirements of ANSI/NISO Z39.48-1992 (Permanence of Paper).

CONTENTS

INTRODUCTION

WHAT IS ECOLOGY?

Above all else, ecology is a science. Ecology is not a philosophy or a way of being, and even less is it a panacea to save the planet. Thus, saying a human action or activity is "ecological" makes as little sense as saying it is mathematical, physical, or chemical.

The science of ecology has existed formally for just over a century, focusing on the interactions between living organisms and their environments. The German zoologist Ernst Haeckel coined the term ecology in 1884 by combining the Greek roots *oikos*, "house," and *logos*, "study." With time, many specialized research areas have developed within the science: physiological ecology; behavioral ecology; the ecology of populations, communities, ecosystems, and landscapes; arctic, boreal, and tropical ecology; limnology; marine ecology; forestry; and fisheries, among others.

In studying the relation between organisms and their environments, ecology has allowed us to see the impact human societies can have on nature and helped us realize we don't live in isolation from the natural environment. Even short, localized actions have repercussions at various spatial and temporal scales. However, the contribution ecologists have made to such understanding is the tip of the iceberg, because the natural world has so many possible interactions that need to be studied, and studied on a planetary scale. Despite this continued need, human society has made some important responses to ecological findings, and the discipline still has much to teach us about solving problems caused by urbanization, human population increase, and globalization. It's in this transfer of knowledge that the common confusion between "ecology" and "environmentalism" has arisen.

By providing a scientific framework, ecology can help us predict the biological consequences of our political and economic choices. Using science and its method of hypothesis testing helps us understand the effects of draining a marsh, clear-cutting an ancient forest, introducing an exotic species, dumping waste into a river, or building a new road. What we do with the knowledge we gain is a societal decision on whether to remedy the situation.

The ecologists who helped write this book all wanted to ensure that their scientific work is better understood. We have used the environment most of us are familiar with: the city. Since an increasing proportion of the human population lives in cities, these may be the only environments many people on our planet know well. Luckily a lot of ecology can be understood there, either through direct encounters with urban plants and animals or by applying general ecological concepts still observable. We want this book to demonstrate how you can experience both in your city. We hope it opens your eyes to the life around you. After all, no one escapes the forces of nature alive—even in the city!

Beatrix Beisner, Christian Messier, and Luc-Alain Giraldeau

Professors of Ecology

Department of Biological Sciences

University of Quebec at Montreal

PART I

Around the House and Garden

1

APPLE AT MY CORE

Our human notion of garbage as old, unusable, or unwanted material does not exist in nature. Instead, living organisms pass the essential materials for life to others in a relay race with no end. Death is not final; it's just a transitory state for what ecologists call organic matter. Let's observe the reincarnation of something you might consider garbage—an apple core discarded in your backyard. Although the word organic is often used these days to refer to healthier food choices, in biology and chemistry it means something quite different. In fact, **organic** simply means material that is living, or once was. Much of our kitchen waste slowly disintegrates into its organic components: molecules containing carbon and hydrogen (see the boxed definitions at the end of this chapter). The rest of our household waste is made up of inorganic molecules composed of other elements such as nitrogen, phosphorus, or iron. Many of these products of disintegration become **nutrients** essential for the growth of **primary producers**, which in most ecosystems are more simply called plants. Primary producers are at the base of all food chains because by using nutrients, sunlight, and water they produce new life that other organisms depend on.

In **decomposition**, complex waste material (like the core of an apple) is converted into simpler forms that are returned to the food chain. Many hardworking organisms—perhaps not surprisingly called decomposers—carry it out. Without decomposers, every plant or animal that has died since the beginning of life on Earth would accumulate around us, leaving no room for new life. In addition to filling the planet with waste, each death would sequester more of the nutrients surviving organisms need for growth, eventually leading to the extinction of all life as they are used up.

Organic matter starts decomposing as soon as the living organism stops protecting itself from decomposer attacks, usually on the death of the organism or one of its parts. When an apple is picked the tree can no longer protect it, and decomposition starts (we keep fruit in the refrigerator to slow down the decomposers). Let's see what happens to the apple core you throw on your backyard compost heap.

PHASE 1: DECOMPOSITION

The first organisms to attack the apple core are the macroscopic decomposers. These include invertebrates (animals without spinal cords), such as millipedes, fly larvae (maggots), and earthworms, that cut the core into smaller bits. Then smaller organisms, like protozoans and tiny worms called nematodes, take over breaking apart the garbage as these bigger decomposers leave.

Decomposing organisms feed on waste such as apple cores to fuel their metabolism—the same reason we eat. Metabolism produces carbon dioxide (CO_2) through **cellular respiration**. In this way these initial decomposers produce both CO_2 (a gas) and solid waste products. This solid excrement is rapidly colonized by microscopic decomposers such as bacteria and fungi that complete the work of deconstruction, creating **humus**, soil high in nutrients and therefore useful for plant growth.

The decomposers haven't yet finished with your apple core. Some proteins, sugars, cellulose, and lignin remain in the humus. These large, complex molecules bind up the nutrients the primary pro-

ducers need. Once again the bacteria and fungi work to break these molecules into their simpler constituent molecules and elements. An important example of this conversion from complex to simple molecules is the decomposition of proteins found in dead plants and animals. Proteins are very large molecules made up of amino acids rich in nitrogen (N). Within the humus, nitrogen is still unusable for plants, since it is caught up in proteins. Decomposers convert these into smaller molecules: urea, ammonia (NH_4^+), and nitrites (NO_2), and finally the form plants most prefer, the nitrates (NO_3). Even though N is the most abundant molecule in the air we breathe (78 percent), plants can take up only the forms found in the soil, so microscopic decomposers are essential in degrading proteins into forms of nitrogen that plants can use to make more proteins.

At this point your apple core has completely disintegrated, transformed into its basic constituents of CO_2 and inorganic nutrients like nitrogen. Now it's ready for the next step.

PHASE 2: RECONSTRUCTION

The element at the base of all life on Earth is carbon (C). In plants we find it principally as cellulose and lignin, the major components of wood, pulp, and bark. Carbon is found mostly in animals' tissues, including fat.

All living organisms need carbon for growth, maintenance, and reproduction. Plants take C directly from the air and convert it to other molecules (using light energy) through **photosynthesis**. Animals take in C by eating organic matter like plants or other animals, then convert it into energy by cellular respiration (the same process the decomposers use to keep growing).

This is the final step in the reincarnation of the apple core, now converted into minuscule molecules of CO_2 and nutrients so that plants in the garden can take it up. If you feed your growing vegetables with compost, parts of the apple will become part of your body when you eat those magnificent home-grown tomatoes and cucumbers.

The number of atoms on Earth has remained more or less the same since the planet was formed. Because they are constantly re-

cycled, some of the atoms in your body may have once belonged to Jurassic dinosaurs, while others might have spent time in the body of Plato or Mozart. Then again, maybe your atoms were part of their forgotten neighbors, so let's not get carried away.

Alice Parkes

Try this experiment. Take six to ten seeds from the same plant species (ideally, all from the same package from your local garden center). Beans or peas are easy to grow and measure. Grow half the seeds in pots with potting soil and the other half in a mixture of half potting soil, half compost. Treat your pots the same in every other way (light, water, temperature). And don't forget to water!

Measure the stems every day for a few weeks and record the figures in a notebook. From these measurements you should be able to estimate growth rates; a simple way is to plot size against time on a graph. If you are patient you may even see seeds forming. Is there any difference in growth rate or number of seeds in plants grown in regular soil and those supplemented with compost? Why?

SOME DEFINITIONS

Organic matter: Material made up of molecules that contain at least both carbon and hydrogen atoms.

Inorganic matter: Material made up of any other types of atoms (elements).

Decomposers: Consumers that reduce complex organic matter to simple molecules using oxygen and producing carbon dioxide gas (e.g., many insects, bacteria, and fungi).

Nutrients: Elements other than carbon and oxygen that are essential for life (N, P, K, Ca, Mg, S, Si, Cl, Fe, B, Mn, Na, Zn, Cu, Ni, Mo).

Primary producers: Organisms that grow by using solar energy (sunlight), water, carbon dioxide, and nutrients (e.g., trees, aquatic plants, algae).

Macroscopic: Visible to the naked eye (larger than 0.02 inch [0.5 mm]).

Microscopic: Not visible without special lenses (smaller than 0.02 inch [0.5 mm]).

Humus: Decomposed organic matter.

Respiration: Oxidization of organic matter using oxygen, releasing carbon dioxide and heat.

2

ARBOREAL AQUEDUCTS

It's noon on a sunny summer day. The thermometer reads 90°F (32°C), and not a drop of rain has fallen in weeks. Heat shimmers above the parked cars. The grass in your yard is turning brown, yet the magnificent maple in your front yard doesn't seem to be suffering. Now that you consider it, all the trees on your block seem immune to the drought and still sport very green foliage.

All organisms need water to survive, and trees are far from an exception. So why do they remain green while the grass turns brown? What **adaptations** did trees evolve to allow them to colonize all parts of the planet, from arid deserts to barren mountains to muddy swamps? Don't forget that for many living organisms, too much water is as much of a problem as too little. Scientists were long baffled about how trees survive in such a wide variety of humidity levels.

The secret lies in the way trees transport water up those tall trunks, from the roots to the leaves.

Without water, there would be no life on Earth, or at least not the kind we know. Certainly there would be no trees. As they grow, trees go through a series of complex physiological processes including germination, photosynthesis, growth, and absorption of nutrients from the soil, and they all take lots of water. In a single summer, a large maple tree transports up to 53 gallons (200 L) of water every hour from its roots to its uppermost leaves.

How do trees pump all this water from the soil to the impressive heights where their leaves are found? For some trees the task seems downright impossible: consider Australian eucalyptus or the California sequoias, which must pump water up 500 feet (150 m).

MODULAR TUBES

Trees constitute a complex network of natural aqueducts. Just like municipal waterworks, arboreal aqueducts must constantly adapt flow to the amount demanded by the end users—in this case, the leaves.

Trees take water from the soil using their smallest roots, called root hairs, but some very small fungi (**mycorrhizae**) that colonize root hairs do most of this work. Mycorrhizae are indispensable to

the survival of most trees: the minuscule filaments (**hyphae**) of the fungi vastly improve the tree's ability to absorb water and nutrients. The hyphae reach into and exploit resources from a much larger volume of soil than could the roots alone, while remaining attached to the tree's root hairs. In fact, the roots, root hairs, and mycorrhizae occupy as much volume as all of the tree's foliage. This extended root system gives trees a major advantage over lawn grasses, whose roots are often confined to the top 4 inches (10 cm) of soil.

After being picked up by the roots, water continues to travel through the plant by small vessels in the **sapwood**: mainly living tissue directly under the bark that makes up the outer 2 to 6 inches (5–15 cm) of the trunk. Acting like a sponge, sapwood moves water toward the leaves. In contrast, **heartwood** is the dead tissue in the middle of the trunk that holds the tree up. At the summit, the sapwood divides and makes its way into each branch and twig to irrigate every leaf, so even the most remote receives the water it needs—most of the time.

THE SECRET PUMP

Now that we better understand the route water takes through the tree, we can look at what makes it move. At one time researchers thought tree roots pushed water from the soil up to the leaves using "root pressure." But they quickly learned that if such a mechanism existed, it could not raise water higher than 10 feet (3 m)—certainly not high enough to reach the tops of most trees. To see how the process works, we need to understand what leaves do.

Leaves transpire. Lots. Plant transpiration is much like human perspiration. The foliage of a single tree produces enough water in a day to fill at least ten bathtubs, without expending metabolic energy, owing to a pressure differential between the atmosphere and the inside of microscopic pores (**stomates**) found on the underside of every leaf. During the day, each leaf dissipates heat by evaporating the water in its cells, creating water vapor within the leaf. Because it is under pressure, this vapor seeks to escape whenever the leaf opens its stomates, as it must do to capture carbon dioxide (CO_2) in the air for photosynthesis. So a leaf picks up CO_2 while releasing water vapor. This simple transfer of water from the leaf to the atmosphere, **evapotranspiration**, causes a chain reaction: to fill the vacuum cre-

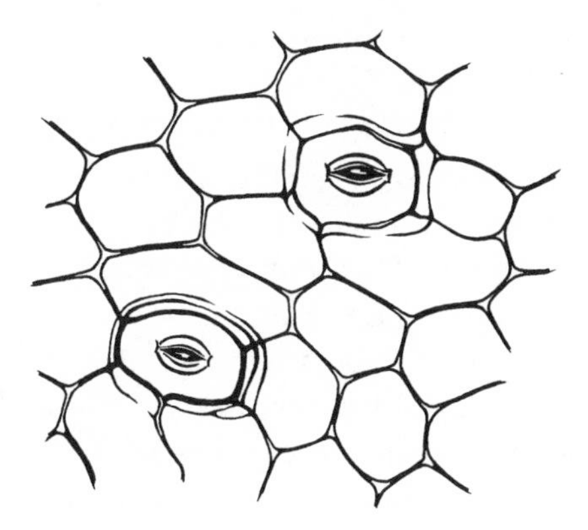

ated by the lost water, leaves suck more water from the trunk, the trunk in turn sucks water from the roots, and finally the roots absorb more water from the soil.

During droughts, urbanites may be asked to not water their lawns, to avoid washing their cars, and sometimes even to limit their time in the shower. Similarly, there are times when the tree needs more water than the environment can supply, making rationing necessary. A tree rations water by partially or completely closing the stomates. But this water saving comes at a cost, since it limits the amount of CO_2 the leaves can take up, thereby reducing photosynthesis and ultimately growth. If the water shortage goes on too long, the tree will lose vitality. The year after a severe drought, it is common to see dead branches near the tops of trees.

Thus a tree faces a trade-off in acquiring resources. To capture enough CO_2 to survive, it must open its stomates. But the more it opens them, the more water it loses. This dilemma applies to most plants except the cacti. Cacti and some other plants from arid regions have developed a different photosynthetic machinery to capture CO_2 at night when it is cooler, minimizing loss of water during the hot days.

OAK EFFICIENCY?

There are two types of trees: those that waste water and those that save it. Wasters are found in wet or humid environments because they cannot survive without plenty of water. These arboreal wasters, including willows, cottonwoods, silver maples, and black ash trees, are generally found near water and in the floodplains of rivers. In these habitats, water wasters rapidly outcompete other species, but they remain vulnerable to prolonged droughts.

At the opposite end of the spectrum are water-saving species like pines, cedars, and oaks as well as species that grow in desertlike conditions such as those in the southwestern United States. These species control the opening of their stomates very precisely to minimize water loss. Their root systems are well developed, efficient, and very deep, so they can search more extensively for water. Their overall growth tends to be slow, but these trees can survive with very little water. They often occur on rocky outcroppings where they are competitively dominant.

In regions subject to occasional droughts, trees should evolve the strategy of dropping their leaves at those times to reduce their water requirements (of course, the trade-off is that they won't photosynthesize). Another strategy for dealing with drought is to produce more roots, maximizing water absorption when demand is high. Trees have evolved many ways to get the water they need. But one thing is certain: a water-wasting tree will never occur naturally in desert regions. Maybe humans could learn a thing or two.

Christian Messier, Sylvain Delagrange, and Frank Berninger

On a sunny summer day, use twist ties to fasten a clear plastic bag around a leaf growing in full sunlight and another around a leaf of the same plant species growing in the shade. Look at them after a few minutes. What differences do you see? Can you relate your observations to what you learned in this chapter?

THE ECOPHYSIOLOGY OF TREES

To better understand how trees have adapted to their environment and quickly colonized so many habitats, scientists have combined two fields of scientific study: physiology and ecology. Physiology explains how small-scale mechanisms work, like the opening and closing of stomates within leaves. Ecology helps explain why particular mechanisms allow a species to survive better in one habitat than in another.

Ecophysiology combines both approaches. It is a young science that uses many tools, including computer models. With growing knowledge about the physiology of trees, researchers have developed more and more precise models for simulating the growth of a virtual tree under a variety of environmental conditions. Using these models, they can explore the effect of giving a tree the leaves of a water waster such as the ash but the structure of a water saver like the oak. Models like these help researchers understand how the various strategies different species adopt help them survive.

3

LAWN LIONS

Finally, a peaceful moment! The neighbors have silenced their lawn mower and are kneeling quietly with trowels, trying to eradicate the source of the luxuriant yellow carpet that covers their lawn each spring. Just how did this small plant, originally from Eurasia, become such a successful invader of lawns throughout the world? As often happens with exotic species, humans have aided the dandelion's success. Perhaps our delight in blowing the seeds off dandelion heads is at least partly responsible.

An **exotic species** is one originating in another region of the world that has successfully moved past some natural boundary like

a mountain range or an ocean to colonize new habitats. Some exotic species cannot survive in their new environments without human aid (purposeful or not). Others adapt so well that they become naturalized.

Among exotic species, some spread so rapidly that they become invasives, coming to dominate their new ecosystems and even eliminate native species. This is why ecologists often regard invasive species as a threat to natural biodiversity.

In general, a species becomes highly invasive only when a new environment offers it several advantages:

1. Adequate environmental (habitat) conditions.
2. A vacant ecological niche: a set of resources not already being used efficiently by native species.
3. An environment that lacks natural predators (including diseases, insects, or animals) that normally control its population in its native habitat.

The dandelion is an exotic species in North America, but it cannot really be considered as threatening biodiversity: it invades only habitats already strongly altered by humans, such as grassy lawns. Though many homeowners see it as undesirable (a weed), this view is more subjective than based on ecological reasoning.

In fact, early European settlers purposely introduced the dandelion (named from the French *dent de lion*, "lion's tooth," because of its serrated leaves) along with several other common plants used in cooking and in medicine, such as clover, mustard, daisies, and wild chicory. In North America, exotic species make up at least a quarter of the diversity of herbaceous plants. But some introductions were accidental, when seeds were stowed away in packing material or in the fodder brought to feed livestock on board ships.

A great many of the plant species introduced to North America originated in the steppes of western Eurasia and the mountains of Europe and prefer sunny, open areas. Such habitats were found naturally in the new countries' prairies, but also in areas the colonists had cleared of forest. Without human help, many of these species would have been less successful at establishing themselves across the continent.

Cutting forests, clearing land for agriculture, and building cities have all helped many exotic plants spread by literally following in

the footsteps of human explorers. In the seventeenth century, some native populations dubbed the common plantain Englishman's foot or white man's foot because it showed up wherever Europeans set up their colonies.

As Europeans settled all regions of North America, exotic species followed the same routes, growing along roadsides and railroad tracks, spread by horses, wagons, and later cars. Invasions are still being triggered by accidental introductions as current human alterations to habitat continue to benefit exotic species.

ATTILA OF THE LAWN

But let's return to the dandelion. Lawn now makes up the largest cultivated surface in North America (surpassing even fields of wheat and corn). Lucky for the dandelion, which feels right at home in short grass. What could be closer to its native steppes? North America has few native species adapted to these open, grassy habitats, which mostly never existed before, so the dandelion has little competition. Furthermore, its morphology and its life history strategies (the ways it reproduces, disperses, and uses resources) make it the ideal invader.

First consider dandelion morphology: the leaves are arranged in a basal rosette, particularly efficient on a lawn of short plants. The rosette lets the dandelion capture a maximum of sunlight (lots of well-spaced leaves) without having to create an elaborate and energetically expensive support system for them (the stem). Further, the plant resists the lawn mower's deadly blades because its leaves are so close to the ground.

The dandelion's deep root, resembling a small carrot, lets the plant penetrate deeper into the soil than the grasses around it, increasing its access to water and nutrients and making it hard to pull up. This tuberlike taproot also allows the plant to stockpile reserves during the summer, giving it a head start on growth the next spring.

Because it quickly produces new flowers whose bright yellow attracts insects to pollinate them, the dandelion can usually set seeds before a homeowner mows the lawn again. The simple stem that supports the flower grows rapidly, allowing the plant to rise above the surrounding grasses and spread its abundant seeds over great distances (up to 6.2 miles [10 km]) when the wind blows. The dandelion is the ultimate vegetative warrior.

THE RESPONSE

So how do you deal with all those dandelions that have taken over your lawn? Probably the most efficient way is to make the habitat around your house less appealing, as different as possible from the Eurasian steppes where the plant originates. Instead of trying to eliminate dandelions from the perfect habitat, why not manage your yard to favor native species and literally shade it out?

If you are determined to have a perfect grassy green lawn, herbicides are not the solution. They harm the environment, and dandelions simply evolve resistance that requires even more toxic herbicides in the future. You're better off in the long run getting down and fighting your enemy with a trowel.

And here's something else to motivate you: Why not include the casualties of your battle on the menu? All parts of the dandelion are edible and are rich in vitamins A and C and iron. The young leaves are nutritious in soups and salads, and you can cook flower buds like asparagus. Dried and roasted, the roots make a hot beverage like the dandelion's cousin chicory. And the flowers can be used for homemade dandelion wine. It goes without saying that putting herbicides on your lawn rules out these delicious uses—so think before you spray.

Isabelle Aubin

Try to identify organisms (plants, insects, birds) that you find around your house or apartment. You can use the Web or, better yet, field guides from the nature section of your local library or bookstore (get ones that match your region). Once you have identified a few species, do some full Web searches to learn more about where they come from. You'll be surprised to learn that more than one out of every four species surrounding you is exotic. What do you think might be the consequences of such a high rate of introduction of new species?

THE LIFE CYCLE OF THE DANDELION

1. A seed arrives floating on the wind, attached to a tiny soft parachute.
2. A seedling develops as a small basal rosette.
3. The leaves continue to grow close to the soil, providing the plant with energy through photosynthesis.
4. A root like a carrot penetrates the soil and accumulates nutrient reserves during the summer, so the plant can start growing quickly the next spring.
5. The composite flower is made up of hundreds of tiny ray florets. Its bright color and generous nectar attract insect pollinators.
6. The flower quickly changes into a globular, downlike seed head packed with seeds that are spread by the wind to start new plants.

4

PRAISE FOR LAZY GARDENERS

You gaze enviously at the impeccable lawn of your neighbor on the left (neighbor A) and perhaps feel a bit inadequate. But you feel somewhat better about your own lawn when you look over at the neglected lawn to the right (neighbor B). You've noticed the disparaging looks neighbor A gives neighbor B's front lawn. How does the ecology of these two patches of urban land differ? How would an ecologist view them?

The perfect grass lawn: Is it as attractive to insects and other animals as to neighbor A and many other urbanites? How does the overall plant production of neighbor A's perfect monoculture differ from a field of wild plants more like neighbor B's yard? These questions are similar to some important ones currently being posed by ecologists: Why is biodiversity important? How does habitat structure affect biodiversity?

An ecologist asking these questions about the two lawns would first determine the amount of biodiversity in each yard. The simplest

way is to count the species of plants and animals in each to obtain a key biodiversity measure called **species richness**. Ecologists have spent a lot of time determining how the number of species influences the functioning of communities and ecosystems in habitats around the world. It has been shown, for example, that the more species-rich a community, the more stable it is—the more likely to function normally after a disturbance.

More particularly, several ecological studies have shown that a diverse community is more resistant to epidemics, extreme temperatures, invasions by exotic species, and other disturbances. This observation has become so widespread that it has its own name: the insurance hypothesis. Having more species in a community is like insuring your home against calamities like fire and theft.

NO-RISK INSURANCE

The insurance hypothesis is thought to work because the more species are present in a community, the higher the odds that one will survive whatever disturbance might arise. Don't forget, however, that these must be native species, not new exotics (see chapters 9 and 23 for the calamities that exotic species can cause in themselves).

Let's apply the insurance hypothesis to our two lawns and think about the climatic conditions they are exposed to over the course of a year. In many parts of North America, lawns experience a huge range of conditions from dry, snow-covered winters to wet springs followed by hot and often dry summers. In other regions they may simply go from a very wet season to a drought. Add to changing seasonal conditions assaults such as being trampled by children playing or "watered" by a passing pooch, and one realizes lawns face an impressive number of stresses. It is difficult to imagine that one species could have all the adaptations necessary to remain green and healthy under all these conditions. In fact, this is precisely why grassy lawns need so much tending with water, fertilizers, fungicides, and pesticides—they are not adapted to all circumstances. In the case of our two neighbors, lawn B is far more likely to thrive with little tending, since it has many more kinds of plants. The effects of various disturbances will be diluted over many species, allowing it to stay green all year (except when covered with snow), albeit with different plants dominating at any single moment.

WOULD THE FAUNA PLEASE RISE?

In addition to conferring greater resistance to weather, children, dogs, and disease, the greater plant diversity on lawn B creates conditions especially favorable to animal life (including insects). Having several types of plants creates a variety of habitats and environmental conditions, leading to a greater diversity of ecological niches.

An **ecological niche** consists of the range of food and habitat requirements a species needs to survive. A lawn made up of different plants creates a range of possible niches, so more species can flourish than in a lawn where a single species (**monoculture**) is maintained.

Take the ladybug. It obtains important protein for growth by feeding on dandelions and clover early in the summer. The ladybug in turn is an important predator of many insect pests, like aphids, that harm your garden. A lawn consisting of a single grass species (or perhaps two) does not provide the ecological conditions the ladybug needs to grow and reproduce, so in such simple habitats it will not survive to provide any benefits to you.

The difference in habitat structure provided by the diverse set of plants on neighbor B's lawn provides habitat heterogeneity to animals living there. Ecologists have shown time and again that more

heterogeneous environments allow more species to coexist, benefiting the ecological integrity and functioning of natural ecosystems. Without a doubt, neighbor A is doing a lot of work for little ecological gain.

Allain Barnett, Richard Vogt, and François Guillemette

Inventory the diversity of insect types on a patch of perfectly maintained lawn a yard square and compare it with a similar patch in a vegetated vacant lot. You're likely to be astonished at the difference.

5

THE EVOLUTION OF FOOD

What could be more pleasant than visiting the farmers' market, where stalls overflow with fruits and vegetables, each more colorful than the last? It is without question the most biologically diverse place in the city. As we shop for the foods we need, it's easy to forget that the vegetables we eat have resulted from millions of years of evolution. On your last shopping trip you likely gathered some important specimens. Let's open the shopping bag.

According to the most recent estimate, plants originated 3 billion years ago, approximately 1.5 billion years after Earth was formed. The very first living cells were made from complex organic molecules, and the plant kingdom arose as these cells used sulfur, nitrogen, iron, and of course sunlight to produce a great variety of compounds.

Chlorophyll is the most widely found protein on the planet. It is the chlorophyll molecule, along with solar energy, carbon dioxide (CO_2), and water, that allows plants to photosynthesize—to make their own food. Thanks to such self-sufficient organisms (**autotrophs**), organisms like humans that rely on eating other organisms (**heterotrophs**) were able to evolve.

The aquatic environment was the cradle of vegetative life, a nutrient soup that gave those first living cells access to essential components. Nonetheless, it took 400 to 500 million years for plants to appear. Plants first adapted to the environment on the young Earth by specializing their various anatomical components. Roots, for example, evolved to provide solid anchorage, in addition to searching out water and nutrients. Stems elongated to let the leaves capture more light. Plants developed such a diverse set of reproductive systems that they are used to classify all of vegetative life, edible or not.

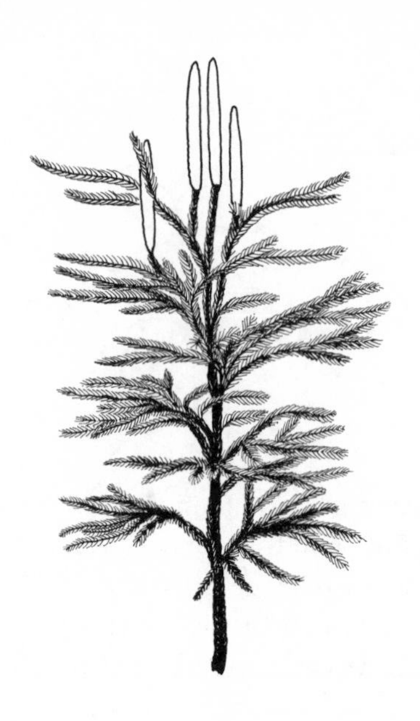

The first terrestrial plants had relatively simple reproductive structures. Present-day forests and flower beds shelter some survivors from this era, including clubmosses (e.g., *Lycopodium*), horsetails, and ferns. Do these ancient species make it into our grocery bags and eventually onto our plates? No, we eat very few of these ancient species—mainly fiddleheads, the springtime shoots of some fern species.

The **gymnosperms**, "plants with naked ovules," so named because their seeds are not protected by an external envelope (Greek *gym* means "naked"), descended from these relatively simple terrestrial plants. Today the gymnosperms are represented by genera and species in the taxonomic group known as conifers (pines, spruces, hemlocks, etc.). This group is not very nutritious, however; if plant evolution had ended there, humans would probably never have existed. Conifers do provide a few treats, including gin (distilled from juniper berries), pine nuts (the seeds of the piñon pine), and spruce beer (which gets its flavor from an extract of spruce needles). Not really the basis of a well-rounded diet. At the beginning of the Cretaceous, about 135 million years ago, the large and very diverse group of plants called the **angiosperms** emerged. In this group the ovules (seeds) are protected by an ovary that eventually becomes the fruit.

With over 300 families and more than 300,000 species, the angiosperms are by far the most numerous plant group, providing almost all our fruits, vegetables, and grains.

Thus most of the plants we eat arose relatively late in evolutionary terms, particularly the herbs and grains, which appeared less than 30 million years ago. Most grains coevolved with humans, who, by cultivating them so extensively and for so long, contributed in a large way to their evolution and distribution.

TAMED PLANTS

We often forget the years of human labor hidden in our food. Our ancestors from all over the world had to patiently domesticate and "improve" many plants. The first were the grains and the legumes, likely because of their high content of protein and carbohydrates. Most fruits and vegetables were cultivated much later, and some varieties have become available very recently.

Most of our plant food originated in three geographic zones: the Mediterranean region, South America, and Asia. Consequently

these plants' evolution is intimately related to the great civilizations of these regions. Nothing happens by chance. Before maritime shipping and commercial exchanges between countries and continents, humans' diets were limited to what they could find nearby. Europeans' colonialism was directly linked to their interest in exotic plants, particularly those that produced the spices.

WHAT ARE WE REALLY EATING?

We eat only a small fraction of plant species, and often only one anatomical part of each edible plant. Rather than remaining helpless victims of herbivores, such as us, plants have developed a number of effective defenses. Some are armed with toxic substances, others are covered with spines or hairs or protected by skin too tough for sensitive palates. Happily, more peaceful plants "accept" being eaten, saving the energy needed to protect themselves for growth and reproduction. Now let's examine a vegetarian menu more closely, using the eyes of an ecologist.

The Fruit

Among plants' morphological structures, one of the most appreciated and nutritious is without doubt the fruit. We will examine why the fruit has been so gastronomically successful in the next chapter, but first we need to clarify a widespread belief. The truth is, there is no scientific basis for the distinction between "fruit" and "vegetable." From a botanical standpoint, tomatoes certainly, but also cucumbers, corn, beans, and peppers, are as much fruits as an apple or a grape. In all cases they are developed, fleshy plant ovaries that protect seeds and are good to eat. So you can remind your children to finish their fruit if they want some fruit for dessert.

Edible Storage Structures

The terrestrial environment offers living organisms a multitude of more or less hostile and often unstable habitats. Given their inability to move away during droughts, many plants have developed structures for storing reserves. Some species store extra resources in a bulge in

their root systems, as carrots, radishes, beets, and turnips do. Others use subterranean tubers like potatoes and sweet potatoes or rhizomes (underground stems) as ginger does. Onions, shallots, and garlic keep their reserves in bulges of their leaves on a plant that has no stem.

Edible Stems

The best examples of edible stems are sugarcane and bamboo hearts. Often mistaken for a stem, the celery plant is actually a rosette of leaves from which we eat the petiole (the leaf stem). This is also so for rhubarb, whose leaves are not edible (an infusion makes a great pesticide). Asparagus also tricks us—it's a young tiller or spear of an underground stem.

Edible Leaves

Lettuce, spinach, Swiss chard, and cabbages are great examples of edible leaves. Plant leaves are also often used as aromatics to embellish dishes and drinks: basil, bay leaves, and mint to name just a few.

Edible Flowers

Cauliflower and broccoli are bouquets of flowers. Capers, usually sold pickled, are the flower buds of the caper tree. Slightly less popu-

lar but just as interesting are the flowers of nasturtiums, violets, and squashes. The more daring might even try eating gladiolus or carnation flowers, all very tasty.

Consider this on your next trip to the market: our rich and diverse diet is directly linked to the incredible forces of evolution at work in plants. There is much more to evolution than descent from apes.

Christian Messier and Julie Poulin

CAFE CONVERSATION

Try to imagine your diet if you could eat only the plants and animals naturally found in your local area. How many fruits or vegetables would be available? What would you miss most?

6

HEY, TAXI!

Most plants remain rooted in the same place for most of their lives, so fruits are their principal way of "moving" to reduce competition between individuals of the same species and enhance access to vital resources across generations. The movement of fruit also allows generations of plants to exchange genetic material and to colonize suitable habitats. Most plants can't count on self-propelled transportation, however; instead, they rely on water, wind, and animals, so their fruit must have appropriate adaptations.

The great diversity of fruit in the produce section of your local grocery store results mostly from how plants evolved to move offspring around—what ecologists call their dispersion modes. Understanding how plants disperse is currently helping ecologists predict how species might move northward with global warming and which ones will be capable of moving. To address these larger questions, we need to know what dispersion modes plants have available.

ANIMAL TRANSPORTATION

Many plants rely on animals to move them around, a strategy called **zoochory**. And some use endozoochory, a strategy to get their fruits (and therefore their seeds) inside the animals so they can travel even farther. These fruits develop seductive properties that animals find hard to resist, like vivid colors and tasty, nutritious flesh. Many mammals, large and small, and some birds love berries. Once inside an animal, the fleshy parts of the fruit are digested, and the seeds are expelled sometime later with the other waste products of digestion (which act as fertilizer). Because digestion takes time, the seeds

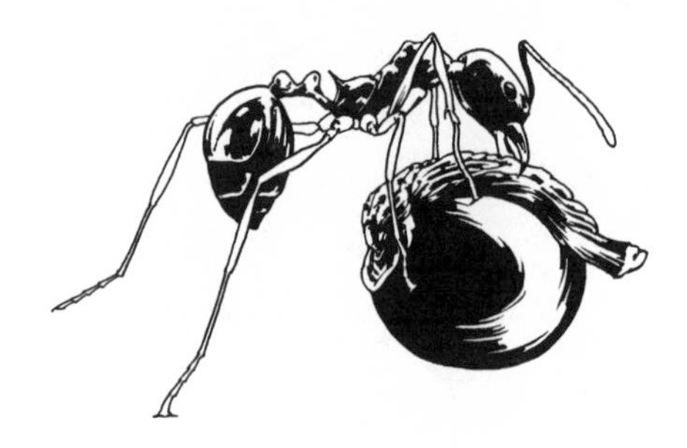

usually end up some distance from where they were ingested. Everyone benefits: dispersion of the plant's offspring is ensured, and the animal gets food. When both parties benefit, ecologists call the interaction a **mutualism**.

Animals do not have to be large to move seeds around. Take ants: they are attracted by the elaiosome, a nutritious fleshy structure attached to the seeds of violets and some other plants. Ants carry these seeds to their underground nests, and when the ant colony eats the elaiosome, the remains are left behind in an ideal place for germination. In fact, dispersion by ants is important for many species of plants and has its own name: **myrmecochory**.

The seeds of certain species—such as the raspberry—germinate better after they have been digested than when they are not eaten at all. On its passage through a bird's digestive tract, for example, the leathery envelope of the seed is scarified, mechanically or chemically, tenderizing it and easing germination. However, not all seeds can withstand passage through a digestive system. Nuts like almonds, hazelnuts, and acorns are very nutritious, but they can't survive digestion. Then why do they make themselves so edible? These plants use a slightly different strategy: the animal carries them away, but not in its stomach. Small animals like squirrels and chipmunks hoard these seeds in very large quantities, and many are forgotten or left untouched if the animal dies or abandons a particular burrow. Once again the seeds find themselves far from where they were produced and, perhaps just as important, often in an environment that promotes germination. So even for appealing and nutritious seeds, the dispersal mode defines the characteristics of the fruit.

The position of the seeds within the fruit's flesh also depends on the animal dispersers, and we can often guess which type of animal

is important for dispersing a particular plant's seeds. Strawberries, for example, are often eaten by very small animals with tiny stomachs that are likely to fill up before the animal has reached the fruit's center. For the strawberry, then, putting the seeds in a central core (as apples do) would not be an efficient way to get them inside their small animal dispersers. Strawberries have evolved to place their seeds on the surface of the fruit so they are ingested first.

For another interesting twist, consider Velcro, a hook-and-loop tape now widespread in daily life, replacing shoelaces, holding up upholstery, used anywhere two surfaces need to be joined solidly but temporarily. Velcro was directly inspired by a particular plant dispersion strategy called epizoochory, in which the seed anchors itself outside an animal, as on a mammal's fur. Look closely at the burrs of the thistles called burdock that will likely cling to your socks or pants next time you walk through a field. You'll see that the points of the fruit form tiny hooks, just like Velcro. By attaching themselves to animals' coats (and even more inconveniently to children's clothing and hair), these seeds travel considerable distances.

A GOOD WIND

Species disseminated by wind use a strategy called **anemochory**. The shapes of the growths or appendages that let seeds fly vary considerably. The small white plumes attached to the dandelion's seeds look and act like parachutes. Maple seeds (maple keys) have a double sail that makes them rotate in flight like helicopter blades. And the seeds of the birch tree are flanked by small winglike growths thought to carry them great distances on strong winds.

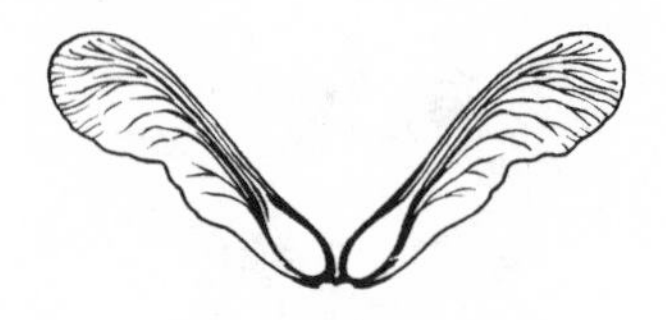

BOMBS AWAY!

You're probably not aware of all the explosions that go on in the quiet understory of the forest. It's an active artillery zone! When ripe, the fruits of the wood sorrel or some species of violets literally blow apart, projecting their seeds up to 16 feet (5 m). Since they do the work of dispersing their seeds themselves, these plants practice **autochory**. You may be familiar with one plant that grows in dry riverbeds and other damp places in more northern climates: the jewelweed, *Impatiens capensis*, is commonly called the "touch-me-not." Perhaps you have seen the small green pods on its bushes late in fall. Finding a ripe fruit that explodes at the slightest touch, propelling its seeds in all directions, is even more fun than popping bubble wrap.

A SOAKER

Living close to water, certain species allow their seeds to float away, a form of dispersion called **hydrochory**. The coconut is probably best known. This nut is enveloped in a thick casing that floats, and it may travel hundreds of miles before landing on a faraway coast (to the joy of island castaways). We regularly see coconuts on the beaches of England, riding the Gulf Stream from the Bahamas. It's a shame that the inclement weather of these northern climates prevents their germination, or we might see palm trees in Brighton.

A HISTORY OF COMPROMISE

As every traveler knows, planning to go to a strange place leads to a trade-off. Often you don't know exactly what to bring, so the more you pack, the better. But the more you carry, the harder it is to get around. Plants encounter the same dilemma. Some species play it safe and invest heavily in individual fruits—like apples, which grow fleshy envelopes to entice animals to eat them. Others, like nut producers, produce large quantities of fruit to give the germinating plant a safe place to grow once it gets somewhere (assuming it's not eaten). For seeds with a more precarious dispersion path, the mother plants may invest in solid, thick coatings (carapaces) for protection. Because constructing these structures takes enormous energy, they produce fewer fruits. By contrast, some species are more adventurous and use the wind to disperse many seeds as fine and light as dust. These seeds are cheap to produce, but they have little baggage and will thrive only if they are lucky enough to fall in the right place. It's like playing the lottery by buying lots of tickets; the winner gets a suitable place to germinate.

Virginie-Arielle Angers and Daniel Kneeshaw

CAFE CONVERSATION

Why do species need to disperse? Humans disperse too. What are the advantages and disadvantages of this behavior for us and other species?

7

SOCIAL LIFE

What looks like a string of three black pearls runs down the street, climbs the wall of your house, and slips into a thin fissure between the bricks. Such intrusions are nothing new; once again a few scout ants have targeted a human abode, searching for resources to sustain their colony or perhaps even setting up a new home—inside yours.

Ant colonies can consist of a few dozen individuals to several million, depending on the species. There are close to 12,000 species on record around the world, probably with many still undiscovered. The size of individual ants varies greatly, from barely a millimeter long to a huge 2.36 inches (6 cm) for queens of the genus *Anomma* from Africa.

TO EACH A TASK

Ants are in the order Hymenoptera, which also includes the bees and wasps. Many species of this order are **social insects** that live in groups. Individuals in ant societies are generally classified into reproductive and sterile castes. Reproductives are winged female queens (or future queens) and winged males. Young queens will mate during the annual swarming period, when all the males and

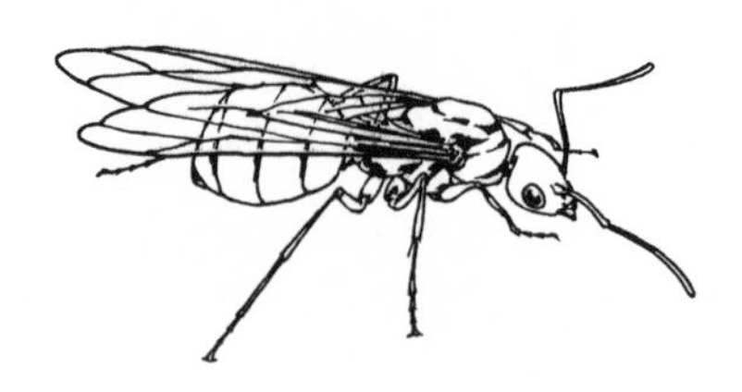

young reproductive females leave the colony together, mating either in flight or on the ground.

Males play a simple role in ant societies: they have no duty but to pass on their genes, so they often die shortly after mating. Depending on the species, a mated queen will either create a new colony or join an established one. Once she finds a suitable egg laying site, she rips off her now useless wings and lives a sedentary life from then on, reabsorbing her flight (alar) muscles to provide extra energy for her next task: laying eggs.

Most of the eggs from the queen's mating flight produce males and sterile females, called workers. These hardworking females take care of all nonreproductive tasks in the nest: caring for the egg brood (the future reproductive females), constructing and maintaining the nest, searching for and collecting food, and—not least important—defending the colony.

Worker ants are the ones running around gardens, parks, and sidewalks. We often see ants coming and going between their anthill and certain plants. If you look at the plants closely you may see

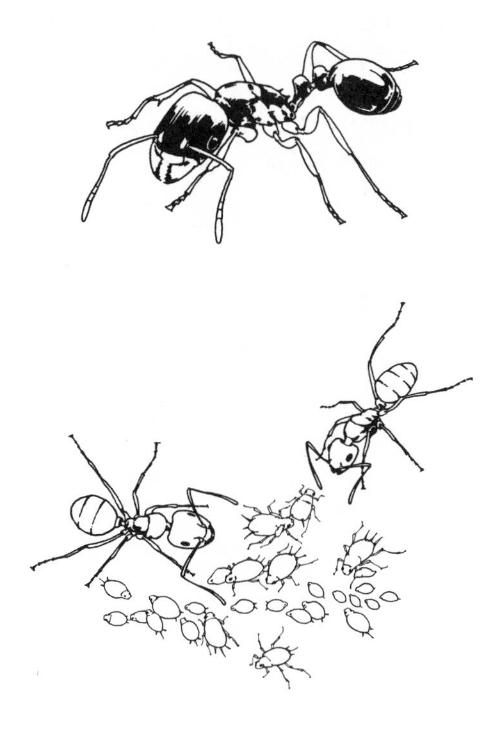

clusters of tiny insects that the ants appear to caress with their antennae: these are aphids, docile livestock that some ant species tend. As they feed on plants, aphids secrete a sweet honeydew that ants love. In return, the ants protect the aphids from predators. As in zoochory (chapter 6), we are dealing with a **mutualism**: an association in which both species benefit.

A ROOM IN TOWN

Ants are expert city dwellers, particularly gifted at taking from others (usually us humans). Even the cleanest of houses, in any season, has something to offer. So, not surprisingly, we may see them running across everything from bathroom floors to the kitchen sink. One of the most common ant inhabitants of houses and public buildings in North America is a small red-brown species, barely 0.08 inch (2 mm) long, called the pharaoh ant (from its scientific name *Monomorium pharaonis*). This species is not native; it seems to have arrived from Africa by way of Europe. In parts of North America where winter takes hold, it has adapted to living in our heated homes, and it will often occupy entire city blocks all year round.

The pharaoh ant differs from our native ant species: it does not have flying individuals that form mating swarms; mating occurs in the nest itself. Another trait that has allowed this exotic species to thrive wherever it travels is extreme flexibility in choosing nest sites: it can use everything from a small fissure in a wall to the bottom of a pot. Even a DVD player can house a colony.

Once one pharaoh ant has been discovered in a house, we can

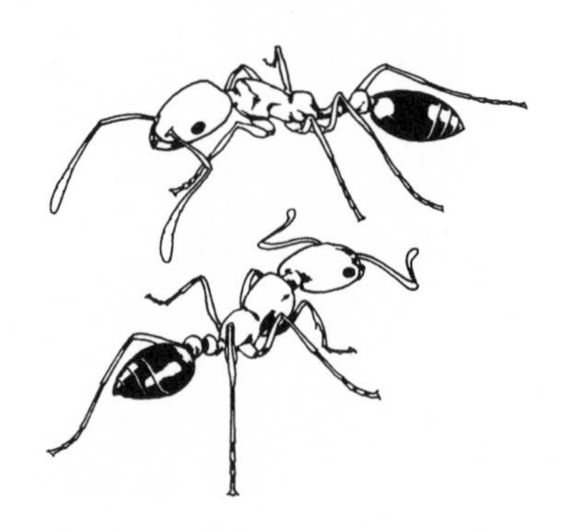

expect to find other members of the colony just about anywhere. If we try to get rid of a nest, the rest are likely to move to a new location within the hour. Because the pharaoh ant eats a variety of items (it is an **omnivore**), it can survive on just about any crumb it finds. But all too often, just as they welcomed themselves into our homes, ants will help themselves to our pantries.

The pharaoh ant is not the only nonnative ant species: North America has over a dozen.

A FIERY INVADER

One of the best-known exotic ants in North America is the red imported fire ant (*Solenopsis invincta*). Originating in South America, they arrived in Mobile, Alabama, in the 1930s and have since spread through all the southern United States. They have a nasty sting that can cause pustules. When threatened they attack in swarms, collectively responding to the release of **pheromones** (chemicals that affect the behavior of other individuals) from the lead attacking ants.

Red imported fire ants are also common in our urban environments, making nests just about anywhere they can, including school yards, parks, and lawns, under sidewalks, next to houses, and even inside electrical equipment like air conditioners and utility boxes. Furthermore, because they form attack swarms, they threaten small animals such as rodents and ground-nesting birds as much as domestic cats do (chapter 12). Perhaps one of their most annoying features is that they build nests in the housings of traffic lights, often shorting them out. These urban ants even get to control traffic!

Benoit Guénard and Amélie Poitras Larivière

CAFE CONVERSATION

Ant colonies divide duties and tasks among specialists, some caring for the young, other gathering food. Do you see a similar division of labor within human societies and corporations? Social insects work ultimately to get their genes into the next generation. If you find a similar division of labor in humans, do you think it arises to accomplish a common objective for all?

EUSOCIALITY: THE MOST HIGHLY EVOLVED FORM OF SOCIALITY

We call many animals social: chimpanzees, penguins, tuna . . . But in reality very few species have evolved the most advanced form: eusociality. Among these are insects including the termites, some wasps, the bees, and of course the ants. Eusociality is defined by four criteria:

1. All juveniles are raised together in a particular part of the colony.
2. Adults cooperate in the care of young: many individuals in the colony, reproductive of not, contribute to feeding, cleaning, and—when necessary—moving the eggs, broods, and nymphs.
3. There are overlapping generations: many generations coexist in the colony.
4. Only some are specialized for reproduction: certain individuals have a reproductive function (queens, or gynes), while others maintain the colonies (workers, or ergates). We speak of different castes that can be differentiated morphologically or physiologically.

There is only one case of eusociality in mammals: the naked mole rat (*Heterocephalus glaber*), found in the desert of eastern Africa. In fact, most animals we call "social" do not have individuals specialized for reproduction and do not satisfy the last criterion for eusociality. Whatever we might think, humans do not qualify as the species in which sociality is the most highly evolved.

8

BATHROOM DRAMA

In the rarely used guest bathroom, a centipede has fallen into the bathtub and feverishly surveys its enamel prison. In its panic, one of its many legs disturbs a thread in a deadly hunter's web. The weaver spider, always on the lookout for prey, immediately strikes out. The centipede, a powerful predator itself, rips a chunk out of the arachnid's web. In a frenzied battle the spider attempts to paralyze the vandal, which seeks to escape at all costs. Despite the lightning speed of the eight-legged beauty, the centipede flees down the drain.

Many animal species, especially domestic ones, live in and around our houses, sometimes permanently. Spiders are one of the most common and most remarkable groups, in both their appearance and their behavior. There are 35,000 species of spiders in the world, and the oldest known fossil dates back some 395 million years.

Since ancient times, spiders have stirred the imagination of humans. The name Arachnida, the taxonomic class spiders belong to, originates in Greek mythology. Arachne was a proud young woman who was an expert weaver. She challenged the goddess Athena to see who could create a more beautiful tapestry. Athena's tapestry

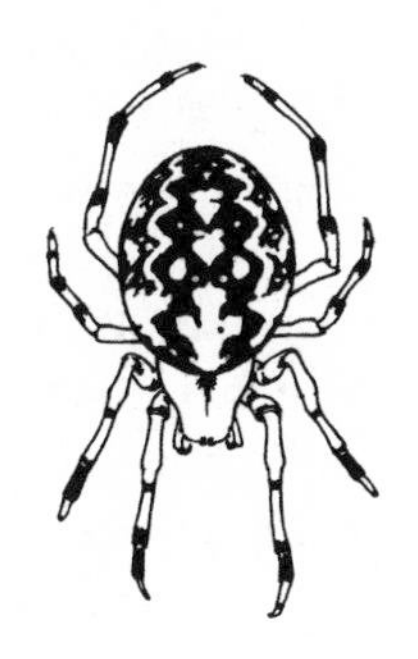

was gorgeous, but Arachne's was just as splendid. Enraged at the mortal's audacity, Athena attacked the young woman, who hanged herself in despair. Filled with remorse, Athena transformed Arachne into a spider so she could weave forever.

From the less poetic viewpoint of taxonomy, spiders are part of one large branch of the phylum Arthropoda: the other branches include the insects, the crustaceans, and the millipedes. The class Arachnida is shared by scorpions, mites, and ticks. The principal characteristics distinguishing arachnids (mainly from insects) are four pairs of legs (instead of three), the absence of wings and antennae, and the fusion of the head and thorax to create the cephalothorax.

LIFE AND DEATH

For the most part spiders are solitary, the exceptions being a few rare social species in South America. To orient themselves, they use four pairs of eyes as well as sensitive vibration-detecting hairs on their legs and specialized appendages called **palpi**. All spiders, whether web spinners or active hunters, remain attached to a silk line while moving, for security should they fall. The life cycle of spiders varies by species, but none undergo metamorphosis as do many insects, most notably the butterflies. Young spiders are mostly just small versions of adults. Generally, females lay their eggs in a silk cocoon that they carry with them or hide in a sheltered spot. After the young emerge, they grow, as many insects do, by **molting**—periodically shedding their skins.

Depending on the species, spiders catch prey either in a web or by hunting. Spiderwebs can take various forms—spiral, platform, horizontal, vertical—and some spiders even trap their prey with a silk lasso. Once captured, the prey is either immobilized with a bite that injects venom or wrapped alive in silk. Spiders that do not construct webs hunt mostly by surveying the ground or branches (or the beams of houses) in search of prey. Once it discovers a prey organism, a spider pounces on it and gives a paralyzing bite. All spiders must liquefy their prey because they cannot chew or absorb solid food. They inject the captured prey with digestive juices, essentially digesting it before eating. Once the spider has sucked out the liquid contents of its prey, nothing remains but an empty envelope.

SPIDER CONTROL

Spiders are voracious predators that can control prey populations under the right conditions. Spiders occur often in houses and usually eat small arthropods like flying or crawling insects (beetles, silverfish, cockroaches, etc.). In nature, they provide a natural biological control for many insect pests.

Spiders are also notorious cannibals. At all stages of development, every individual risks becoming prey to another member of its own species. Spiders even commit matricide—the juveniles of some species, notably the wolf spider, devour their mothers. They can also display sexual cannibalism, in which the female eats the male just after mating; he is really just an appetizer, because males are much smaller than their female partners. These habits might seem strange, but keep in mind that evolution always selects for individuals whose behaviors most favor their survival, increasing their **fitness** to reproduce. Juveniles that eat their mothers obtain the food needed at the start of life, increasing their chances to survive. The mother spider's consolation for this ingratitude is that her offspring are more likely to perpetuate her genes. Male spiders, with little chance of finding another female to mate with, have been selected to provide their own bodies as nourishment for the young embryos their sperm is about to father. What a way to give your all for your children.

When spiders are not hunting each other or their more usual prey arthropods, they hunt other predators they normally compete with. Consider the battle between the centipede and the spider. Both species normally exploit the same prey arthropods, such as silverfish. When a spider captures and kills a centipede, it is eating a member of the **guild** we could call predators of silverfish. A guild is a group of species that all naturally prey on the same species. By eating other predators that share the same prey—**intraguild predation**—the

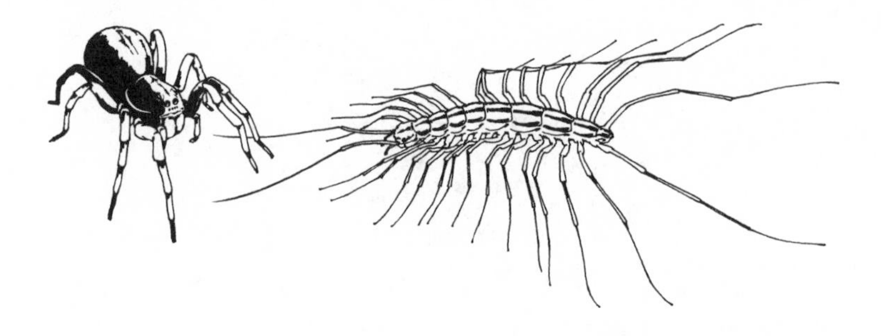

spider not only obtains a meal rich in protein but eliminates both a competitor for food and a potential attacker (the centipede could turn the tables and eat the spider). Without question, this type of predation is risky.

UNWANTED GUESTS?

Spiders live with us year-round, in our basements, our attics, and our bedrooms. Black, furry and fast, spiders are without doubt the most feared arthropods, despite their killing so many nuisance insects and mites. Spiders do not generally bite humans unless they are cornered between skin and clothing or in bedding. A spider bite usually doesn't hurt for long. Rarely, there are allergic reactions, not to the venom, but to the proteins in spiders' saliva. In most cases spiders are completely harmless to humans, and in North America encounters with species that can kill a human are rare. In 1983 an American study revealed that 80 percent of symptoms assumed to be caused by spider bites came from the bites of other arthropods. So don't be too quick to blame a spider.

If spiders are still unwanted guests, you can vacuum up their webs. Blocking access to the house by filling in holes and screening windows will help keep numbers down. When you do find a live spider, it's best to gently encourage it to leave with a broom or, better yet, capture it in a container and let it go outside without killing it. If you find a spider on your body, try to gently push it off rather than squashing it. Keep in mind that these important predators rid our homes of other unwanted arthropods.

Olivier Aubry and Eric Lucas

CAFE CONVERSATION

Do you think it makes sense for people to be more afraid of spiders, centipedes, and snakes than of trucks and traffic? Do you think human babies are born afraid of such animals, or do we simply acquire these aversions from watching others? Evolutionarily speaking, why is it easier to teach a child to fear spiders than, say, cars and toasters, which certainly injure more of them?

9

WINTER WARMTH

With snowstorms, sleet, slush, and freezing temperatures, winter can make life challenging to say the least. Luckily for northerners, homes are designed to offer a refuge against this annual onslaught. Many northern animal species share our desire for comfort in the winter. Take the way many of us have, for several years now, had the "honor" of sharing our houses with the Asian ladybug.

The ladybug (or ladybird beetle) in question, *Harmonia axyridis*, is 0.19 to 0.22 inch (4.8–5.5 mm) long. It ranges from yellow to dark red, and it can have up to twenty black spots on its wing covers, or in some cases no spots at all. Scientists have observed more than a hundred combinations of colors and spots so far, leading some to believe it is actually many species, not just one. In North America the orange variety with nineteen black spots is most common. This beetle can usually be recognized by a pattern of spots that form an M (or a W) just behind its head.

As its name indicates, this ladybug originated in Asia. It was introduced to California as a biological control against aphids first in 1916, then again in the 1960s. Many states followed California's example, mostly between 1978 and 1982. Most introductions were unsuccessful at that time, but since the start of the 1990s the Asian ladybug's range has spanned the continent from west to east. It is impossible to know whether these are all descendants of the original introduced populations or whether some have arrived subsequently, hidden on transoceanic ships entering through various ports.

The population growth of an exotic species in a new environment is often slow, either because species are not adapted to the new climate or because they cannot withstand strong competition from native species. Often, however, repeated introductions will allow

populations that would otherwise persist only at very low densities to increase their genetic variability, permitting the evolution of adaptations favorable to the new environment through **natural selection**: individuals that survive best will come to dominate. In the case of the Asian ladybug, it remains untested whether such adaptations allowed them to flourish or whether they were already well suited to North American environments.

Not only is the Asian ladybug now found across the continent, it occurs both in the countryside and in cities. The reasons it was first introduced as a biological control help explain its success: it is a voracious predator with a large appetite for aphids. It is also extremely fecund (one female can lay up to 3,800 eggs in a year). Unfortunately the Asian ladybug is not content to eat only aphids; it also eats the eggs and larvae of native ladybug species. This **interference competition** is a form of the intraguild predation we saw in chapter 8. Interfering with native species is a common characteristic of highly successful invasives.

Exotic species often have particular ecological characteristics that indigenous species do not. Or exotics may lack natural enemies.

Both these features give them an "unfair" advantage in colonizing new habitats and appropriating resources. The consequences of the Asian ladybug's great fecundity and appetite have been to diminish the abundance (sometimes to exclusion) of several native ladybugs. It also attacks other native insects including butterflies, flies, and other small species, usually easy prey because their vulnerable juvenile stages are found on the same plants where this insatiable ladybug searches for aphids.

SHELTER FROM THE STORM . . .

In their native habitat in Asia, several hundred ladybugs will hibernate together in a rock cavity, sheltered from the wind and cold. Having accumulated fat reserves at the end of the summer, they can survive temperatures as low as 10°F (−12°C). Because many parts of North America have winters colder than this, only individuals that hibernate inside houses will survive. Through natural selection, the number of individuals likely to overwinter in our houses increases exponentially with time (see chapter 20).

As a result, on sunny autumn days we see these ladybugs gathering by the hundreds on the sun-warmed walls of houses, searching for entry points through the cracks around windows or through ventilators. This particular behavior of hibernating in houses is the decisive reason the North American invasion by the Asian ladybug has been so successful.

. . . AND OTHER SCOURGES

By overwintering inside our houses, the Asian ladybug not only avoids lethal cold but avoids drowning in outdoor hibernation sites during the spring thaw. It also escapes many natural predators that benefit from winter aggregations outdoors: parasitoids and entomopathogenic fungi.

Parasitoids are usually small wasps that lay their eggs inside ladybugs. Over the winter, the wasp larvae mature within the ladybug's body, feeding off it until spring. Each larva then forms a cocoon from which a new wasp emerges through the ladybug's cuticle (skin). The

parasitoid completes the cycle by going on to lay eggs inside another adult ladybug the next summer or fall. This parasitism can be an important cause of mortality, especially when ladybugs occur in large aggregations that parasitoids can find more easily. For some species of ladybugs, the mortality rate from parasitism can be as high as 70 percent. For Asian ladybugs parasitism does not reach more than about 5 percent, mainly because our houses protect them.

The other natural enemy of ladybugs in their hibernation sites is the **entomopathogenic** (insect-attacking) **fungus** *Beauvaria bassiana*. When the environment is moist, as it is outside in spring, the fungus attacks ladybugs mercilessly. Because humidity in our homes is usually too low for the deadly fungus to develop, the ladybug benefits.

Thus it is largely thanks to our hospitality that the Asian ladybug wakes up healthy in the spring after a peaceful hibernation. Beware, little aphids—waking ladybugs are as hungry as wolves.

Geneviève Labrie and Caroline Provost

CAFE CONVERSATION

There are two main obstacles to the invasion of exotic species: harsh environments and habitats with few ecological niches available. Exotics must overcome difficulties, whether these are harsh conditions or competitors, predators, and parasitoids. Discuss the role various urban dwellings might play in the success of exotic species.

PART II

In the Neighborhood

10

THE SECRET LIFE OF PONDS

In a park in the middle of the city, a pond offers the urban explorer all sorts of interesting sights. Life abounds: you will likely see dragonflies and other insects flit here and there, and you might see aquatic plants, ducks, frogs, fish, and maybe even a turtle or two. But even if you searched hard all day, you probably wouldn't see most of its biodiversity, because ponds, like all other bodies of water, are inhabited by many microscopic organisms that aren't easily seen with the naked eye.

Most microscopic organisms in our biosphere live suspended in water, often unable to move on their own. This large group is called **plankton**, from the Greek *planktos*, "wanderer," because they are so subject to the vagaries of water movements. Despite their size, plankton are generally abundant (up to several thousand individuals in a quart of water). They are primarily plant (**phytoplankton**) or animal species (**zooplankton**). Without them life in ponds, lakes, oceans, and dare we say the entire planet would be impossible.

PHYTOPLANKTON: EXTRAORDINARY OXYGEN PRODUCERS

Phytoplankton are microscopic algae at the base of the aquatic **food chain**. They are grazed on by the zooplankton, which are in turn eaten by small fish, which are prey for bigger fish, and so on. Not a new story for most of us. But phytoplankton are more than the first link in the food chain; they also photosynthesize, taking up carbon dioxide (CO_2), converting it to sugar using the energy of sunlight, and producing the crucial oxygen molecule (O_2) as a by-product

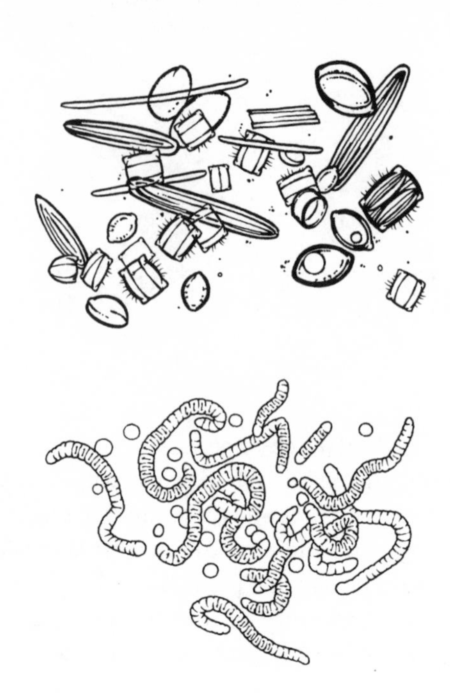

(see also chapters 1 and 2). Aquatic animals like fish and zooplankton must breathe oxygen dissolved in the water. But the importance of photosynthesis goes further. Oxygen in Earth's atmosphere results from the photosynthesis of a very important group of microalgae: the cyanobacteria (blue-green algae).

Phytoplankton come in varied and often beautiful shapes: needles, disks, angular boxes, chains of what look like pearls, spiky spindles, and many more. They are usually single cells, but some species also form colonies. Despite their name, many can move a little: some use air sacs to alter their position in the water column, while others have whiplike hairs or **flagella** that permit rudimentary swimming. Since they are relatively immobile and slightly heavier than water, however, all phytoplankton face the problem of staying suspended in the water column as they must do to gain access to sunlight for photosynthesis to produce food, allowing their populations to grow and produce **biomass** (mass of living material). Because they are at the base of the food chain, they are the main primary producers in aquatic ecosystems, fulfilling the same role that plants do on land (in terrestrial ecosystems). To understand how the phytoplankton biomass reaches the rest of the food chain (especially the fish we eat), we need to examine what happens at the next link (**trophic level**) in the chain. The organisms making up the next trophic level are called the **primary consumers**. In both terrestrial and aquatic ecosystems they are also called **herbivores**, since they eat plants.

THE ZOOPLANKTON: MINIANIMALS

Zooplankton are the main group of animals small enough to locate and pick out the biomass of microscopic phytoplankton that primary producers manufacture through photosynthesis. Individual zooplankton are mostly too small to see easily with the naked eye, but with practice and patience, a trained observer can detect them moving at the edges of a pond. To determine which species are present, an ecologist uses a microscope to see their body parts better. These multicelled animals have simple eyes, several pairs of legs (usually five), a mouth, and a simple digestive system. Evolutionarily they are related to lobsters, crabs, and shrimp, all crustaceans. These animal plankton capture their phytoplankton prey mainly by using their

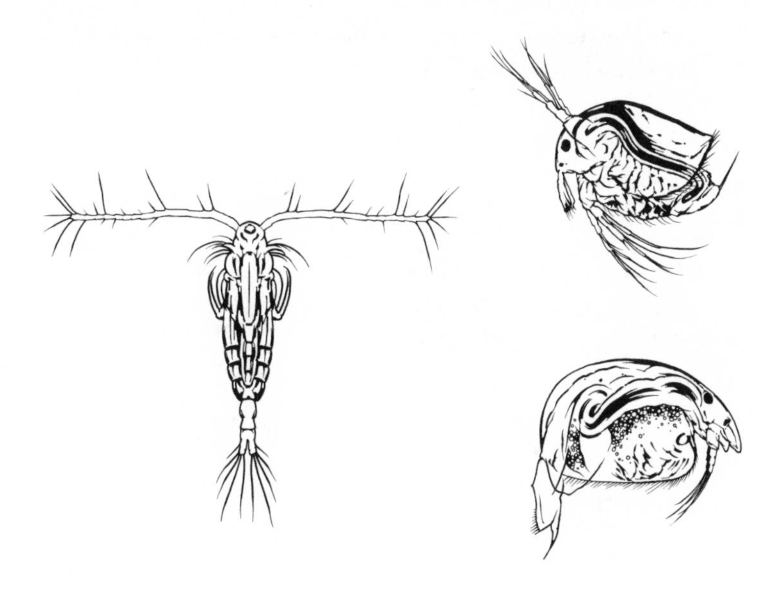

legs to create a small current that flows past their mouths and filtering or picking out the prey cells as they pass. Because they have several sets of legs, zooplankton can all swim and again defy their Greek name.

Zooplankton form the trophic level between phytoplankton and fish. They therefore not only must be good at filtering phytoplankton but must avoid becoming fish food. First, they have evolved transparent bodies, making them harder to see, especially in dark water or at night. Second, they hide during the day, either in the leafy plants at the edges of ponds and lakes or in the deep, dark water of lakes, swimming up only at night to feed on phytoplankton. A third and by no means less important strategy for avoiding predation is **phenotypic plasticity**: an individual can change shape within its lifetime—like a shape changer in a fantasy novel. Although this sounds magical, it's really not that far-fetched. Like all crustaceans, including lobsters and crabs, zooplankton periodically molt by shedding their outer layer (carapace) and growing a larger one. They can change not only the size of their carapace, but some aspects of its shape. Because small fish predators have small mouths, the zooplankton will grow extra long tails or long, pointy heads, making them longer and spikier so they are hard for a small fish to eat. But because growing these features requires extra energy (and therefore food), zooplankton species resort to phenotypic plasticity only when many predators are present.

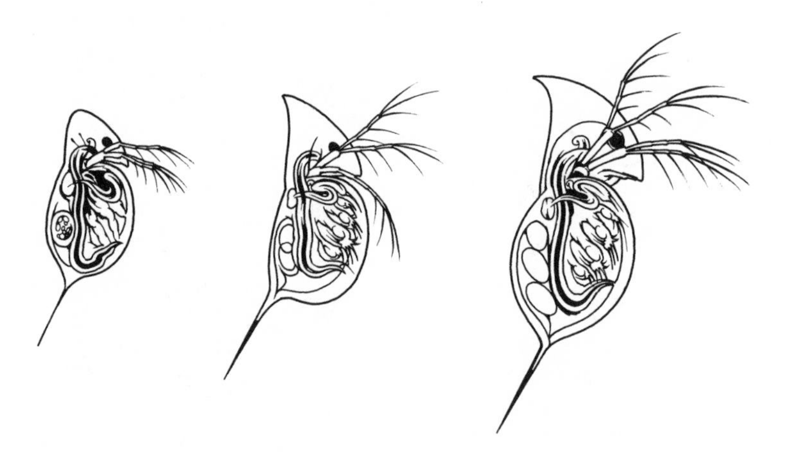

BACTERIA: CLOSING THE LOOP

The remaining major microscopic group inhabiting ponds and lakes are the ones that carry out decomposition, ensuring that dead organisms don't pile up (see chapter 1). Although plants originally create biomass using sunlight, producing food molecules takes more than just light. For the cycle of life to continue, there must be regeneration of other essential elements like nitrogen, phosphorus, and other minerals. Some of the smallest microorganisms, the **bacteria**, carry out this regeneration through decomposition. Bacteria recycle organic matter (usually dead plankton, fish, and plants in aquatic ecosystems). By feeding on the dead material, they separate the original constituents and release them back into the environment so the phytoplankton can reuse the molecules for photosynthesis. In the end, bacteria turn food chains into recycling loops.

Maria Lorena Longhi and Beatrix Beisner

Type "zooplankton" or "phytoplankton" into an images search engine on the Web. Note the wide variety of organisms you see. See if you can find an example of phenotypic plasticity in some species of the zooplankton called *Daphnia*.

11

A LIFE OF EXTREMES

Tree and ground squirrels (family Sciuridae) are common rodent residents of our city streets and parks. They also provide interesting examples of strategies small animals use to deal with extreme temperatures. What do sciurids do in cold weather or in hot desert climates? Have you noticed that some, like the gray and red squirrels, can be seen all winter long while the chipmunk and the common marmot (woodchuck) vanish? Let the inquiry begin!

North America is a continent of extreme temperatures. Winter in the northern regions and summer in the southern United States are challenging for most of us. For small animals that must maintain a constant body temperature as we do (**homeotherms**), a significant problem arises. Let's start with one extreme: a squirrel's approach to winter. Maintaining a warm body throughout cold winters takes lots of energy at just the time of year when food, the fuel for this heat, is least plentiful.

In this ecological context, species have evolved behavioral, physiological, and demographic tactics for survival. Ecologists call these

tactics life history strategies. Each is strongly associated with both the physiological capacity of the animal and its ecology.

Even species that are ecologically similar or closely related through evolution can use very different strategies to survive winter. This will become more obvious when we examine the habits of four cousin Sciuridae species. In case the suspense is too much, don't worry, eastern chipmunks and woodchucks don't really vanish in the winter; it's just that their life history strategy (hibernation) keeps them hidden.

SAVING LIKE A SQUIRREL

Have you ever longed to spend the entire winter inside, eating and sleeping in a cozy chair and never stepping out in the cold? The eastern chipmunk, which ranges from northeastern Louisiana to eastern North Dakota and into south-central Canada, does this every year, especially in northern climates. The Latin name for the eastern chipmunk is *Tamias striatus*, from *tamias*, "one who stores"—a fitting genus for a rodent that spends the late summer stockpiling acorns, beechnuts, maple keys, and other such seeds and nuts in its burrow. The black and white stripes on its reddish brown back give this chipmunk its species name *striatus*, "striped."

In more northern regions, with the first signs of the approaching winter sometime in October, chipmunks take to their burrows and generally don't emerge until April, except for a brief breeding escapade in February that neither snow, sleet, nor rain can stop. They do not spend all their burrow time in deep sleep, however. They alternate periods of lethargy with bursts of activity devoted mostly to feasting on nuts and seeds.

In their burrows, about 1.6 feet (50 cm) underground and well sheltered from extreme variations in temperature, chipmunks enter **torpor**, a deep sleep or "suspended animation" characterized by a drop in body temperature from 98.6°F to 39.2°F (37°C to 4°C) and a reduction in heart rate from 350 beats per minute to a mere four beats. Periods of torpor can last up to six days, followed by very short waking periods when the chipmunk cleans itself and eats some of its stored provisions.

These times of wakefulness help repair the damage caused by

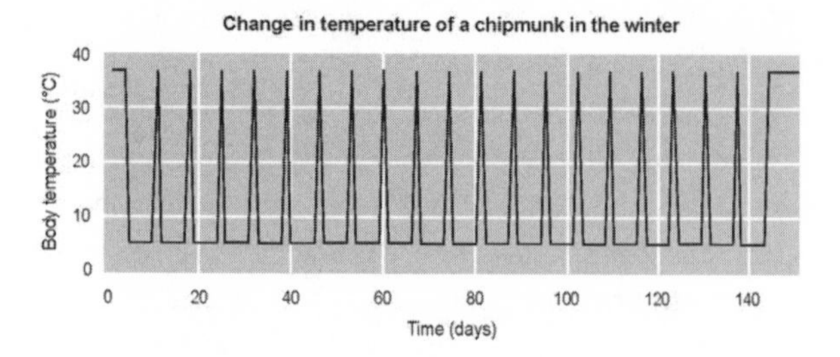

torpor, which is hard on the body, leading to oxidative stress (cellular degradation caused by free radicals; see chapter 22), weakened immunity (increased susceptibility to disease), and sometimes even nerve damage. Hibernation is thus not a completely cost-free solution to surviving winter. It represents a trade-off between the benefits of saving energy when food is scarce and the physiological damage torpor causes.

PUDGY COUSIN MARMOT

The common marmot (*Marmota monax*), also known as the groundhog or woodchuck, ranges from northeastern Alabama and Georgia in the continental United States up into eastern Canada, then spreads north and west all the way to Alaska. It spends winter **hibernating** in its burrow. In contrast to the chipmunk, however, marmots store the food they find in the autumn as fat instead of in a burrow. In the fall, most of what they eat is converted to brown fat, which accumulates under the skin and is especially good fuel to generate for body heat. Human newborns, which are about the size of a marmot, also have lots of brown fat to help keep their small bodies warm. Brown fat can make up half a marmot's body mass at the onset of winter, and almost a third of it will be used (metabolized) during the long winter hibernation, providing a stable and reliable heat supply.

Even if the marmot gets really fat, it can never store enough energy to keep its body at its usual temperature for the whole winter. For this strategy to work, the marmot must turn down its thermostat and let its body temperature decline, much like torpor in chipmunks. However, because their energy reserves are internal, marmots don't need to wake up periodically to eat. So they get fat, then hibernate

in one long episode from which they emerge skinny and hungry in the spring. Because chipmunks stay lean and accumulate reserves outside their bodies, they have to refuel periodically, forcing them into short bouts of torpor dotted with periods of activity. Why have such different life history strategies evolved in closely related species dealing with rather similar winter problems?

The common marmot's way of storing energy (fat instead of food) can be attributed to its food. Marmots eat grass and other herbaceous plants, including their favorites, clover, buttercups, dandelions, and plantains. These food types provide much less energy per unit of biomass than the oily nuts and seeds that chipmunks eat. Thus, while a chipmunk can stockpile enough nuts for the entire winter in some convenient location, storing the equivalent amount of energy in grass would require a ridiculous amount of space and time. In fact it would take a large cave (more than 45.9 cubic feet [1.3 m^3]) to hold the volume of plants a marmot would need to survive the winter: equivalent to a salad of about 250 heads of lettuce. So the animals' diets have repercussions all the way down to their life history strategies and modes of hibernation.

SQUIRREL WINTER SPORTS . . .

Gray squirrels (*Sciurus carolinensis*) range from the Gulf of Mexico east of the Mississippi all the way up to southern Canada, while the range of red squirrels (*Tamiasciurus hudsonicus*) is more northerly, from northern Virginia west to Illinois and Indiana and much farther north into Canada, all the way to Alaska. Unlike eastern chipmunks or groundhogs, gray and red squirrels are arboreal, living in trees. Unlike their ground-living cousins, they do not hibernate, nor

do they enter torpor even in the most northern parts of their range. Instead they spend the winter several yards above the ground in nests, or dreys, made of dead leaves and twigs or in tree cavities. By growing thicker fur and increasing their fat reserves, they insulate their bodies and keep warm. This allows these squirrels to reduce the energy expenditure it takes to maintain their body temperature, known as **thermoregulation**.

Because they do not enter torpor, these squirrels need much more food during the winter than marmots or chipmunks. And because their reserves are external, they have to accumulate a great deal, which is possible because, like chipmunks, they feed on oily nuts and seeds rather that grass and leaves. Nonetheless, gray and red squirrels have to be energy efficient if they want to make it through the winter on their food caches, so they are far less active in winter than in summer. Even more, they restrict their time outside the nest to the warmest time of day. When they do leave their nests, squirrels dig through the snow searching for the caches of seeds they stockpiled in fall in the humus and dead leaves close by. In the city, they also solicit food from passers-by who hand out bits of bread, peanuts, and seeds. And of course they take advantage of bird lovers, who set up bird feeders in their backyards only to end up feeding legions of persistent, bold, and ingenious squirrels . . . and—oh, yes—on occasion some birds.

. . . AND DESERT SPORTS

We have seen how squirrels deal with extreme cold. But how do animals like the Mohave ground squirrel (*Spermophilus mohavensis*) that lives in the Mojave Desert of California deal with extreme heat? The story is not that different. These medium-sized squirrels (8.3 to 9.1 inches [210–30 mm] long), are light gray to brown with cream-colored bellies. Because they also feed primarily on the leaves of plants, they take an approach to the heat similar to the way their northern marmot cousins deal with cold. Mohave ground squirrels **estivate**, which is like summer hibernation. When air temperatures get above 98°F (37°C), usually sometime between June and September, these squirrels enter their burrows and sleep for up to seven months, surviving primarily on accumulated fat. So very different extreme temperatures, but the same feeding constraints, can provoke similar escape tactics.

IMPORTANT TRAITS

No animal can do it all when faced with extreme temperatures; each species is limited by physiological, genetic, mechanical, and ecological constraints. Life history traits or strategies are measurable characteristics of living organisms that are directly related to their reproduction and survival. The most important traits, and those most commonly measured by ecologists, include size at birth, growth patterns, size at maturity, number, size, and sex ratio of offspring, reproductive investment as a function of age and size, and life span. The complex interaction of these traits, characteristic of each species, defines their life history strategies and helps explain, for example, why they do what they do to survive harsh winters.

The next time you notice a life history trait for an animal or a plant species and wonder how it might have come to be, remind yourself that you are in good company. Somewhere out there an ecologist is undoubtedly asking the same question.

Julien Martin

CAFE CONVERSATION

Compare the life history traits of current humans living in the Western world with those you imagine for our cave-dwelling ancestors who had a thirty-year life expectancy. What would have changed?

12

PROWLING PREDATOR

She is everywhere. Hiding in the shadows or patrolling her territory, senses on full alert as she searches out her next victim. The urban jungle is the playground of this dangerous carnivore. She has lethal claws, sharp teeth, and keen eyes. Her agility is legendary. Tremble, small rodents! Fly away, carefree birds! The domestic cat is on the prowl. The battle is heavily weighted in her favor, because she has a powerful ally—humans.

Predator and prey populations have an undeniable demographic link, well illustrated by the synchronous oscillations of the lynx and hare populations observed for over a century in the North American Arctic. Here's how it works: when there are many hares, the lynx get enough to eat and reproduce, and their young can survive and grow. As the population of lynx increases, they exert more **predation pressure** on the hare population, and after some time it declines significantly. Now, because hares are less common, it is harder for lynx to find food. Their weakened state subjects them to disease and famine, and their numbers decline. Predation pressure thus decreases, allowing hares to increase, later followed by the lynx population. The cycle starts anew and continues ad infinitum.

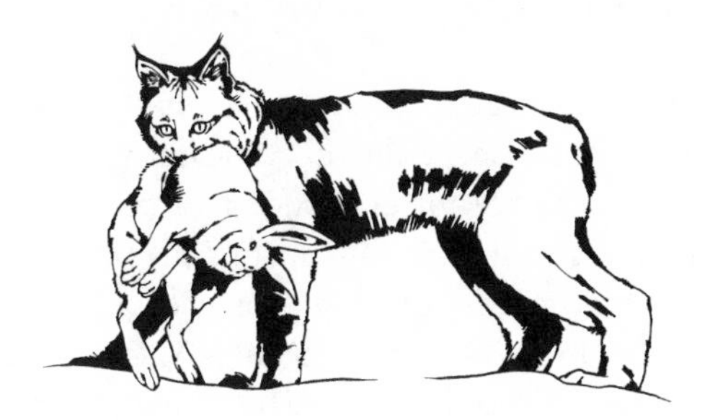

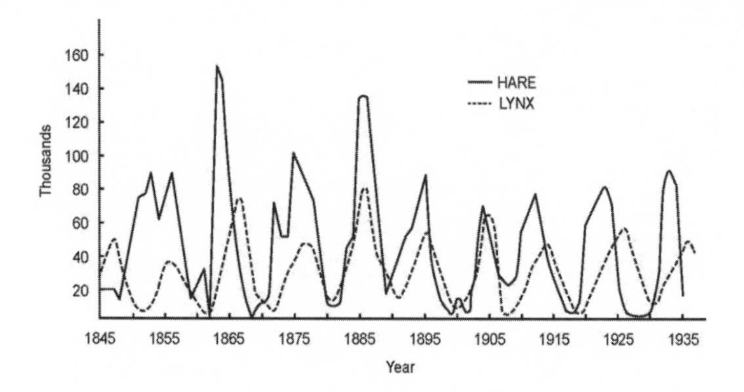

To summarize: fewer prey means fewer predators; fewer predators means more prey; more prey means more predators; more predators means less prey; and so on. Predation not only feeds the predators but can prevent destruction when the environment is overrun by prey (e.g., complete removal of vegetation by herbivores). Plants are protected from hares when lynx predation is high, but if prey become too rare, predators starve.

The demographic link between these two species results from the parallel evolution of predators and prey in the same ecosystem. Too often, disrupting this relationship has catastrophic consequences. For example, the extermination of wolves enabled a demographic explosion of deer across all of eastern North America, with loss of vegetation. In contrast, when mongooses were introduced to Barbados, the population of egg-laying tortoises was reduced to critical lows. Unfortunately, a similar effect occurs in the alleys and backyards of our cities, where the domestic cat reigns.

ORIGINS OF AN URBAN PREDATOR

The Egyptians domesticated cats six millennia ago, and the Greeks introduced them to Europe. From that time until very recently, cats were appreciated mainly for eradicating vermin. The reduction in the population of nuisance rodents (mainly rats and mice) has changed the cat's primary purpose from mouse trap with paws to the more comfortable job of human companion. These days the domestic cat is regularly fed abundantly from the household pantry and no

longer has to rely on the less predictable spoils of hunting. Unfortunately, however, scrawny or plump, a cat will always hunt.

AN APPETITE FOR BIRDS

Researchers recently followed the hunting behavior of 1,000 cats in Great Britain. These cats caught more than 14,000 small animals in a single year (rodents and birds). Summed across all the domestic cats in England, this is equivalent to 275 million prey killed each year.

In North America, almost every third household has a cat. But more worrisome, there are even more stray cats on our streets. Most house cats are in urban areas, so at the continental scale cats kill several hundreds of millions of small mammals and birds every year.

Feline predation is potentially catastrophic when we consider that the domestic cat is an introduced exotic species, and native mammals and birds have little or no defense against it. But the most insidious aspect is without a doubt the cat's tacit alliance with humans, which gives it a considerable advantage over its prey.

SUPPORT FOR WILD ANIMALS

Because they are fed and protected, domestic cats escape the demographic cycle described earlier (see also chapter 20). If they were truly wild, cats' population would normally decline following overconsumption and a reduction in their prey populations. Instead, cats escape this natural adjustment because they have alternative food sources that don't fluctuate. With no regulation mechanism in place, the cat has the power to lay waste its prey populations.

Cats can reach high densities because they are protected against hunger, disease, predation, competition, and the cold. Even if the number of prey in its neighborhood declines considerably, the average cat has only to come home to find a bowl of savory kibble. Domestic cats thus escape the nasty consequences of running out of food: increased fighting with other predators, disease, famine, and often death.

Seeing the multitude of birds visiting their backyards, some people might argue that the situation is not that worrisome. But consider which bird species are perched on our feeders. In most cities these are mainly introduced or exotic species (pigeons, house sparrows, starlings), with little representation by indigenous species (native sparrows and other passerines). Studies in the United States have shown that the more domestic cats present, the fewer indigenous species of rodents and birds. Furthermore, where the domestic cat reigns, exotic species move in and replace the hunted indig-

enous ones. Birds that feed or nest primarily on or near the ground are among the most affected. In the long term, cat predation could drive several native rodents and birds to extinction.

SAVE THE BIRDS?

Cats are an integral part of many people's lives (even for birders), and they are undeniably a part of our urban ecosystem. Cats are here to stay, and that's fine. But there are solutions that could reduce their negative impact on urban fauna. The simplest is to keep kittens inside and give them toys they can "hunt" around the house so they grow up as house cats. As they grow older, these indoor cats will likely be uninterested in real hunting or even in going outside.

Those whose adult cats are accustomed to the freedom of the outdoors need not despair. Cats with the right temperament may be gradually turned into living room predators by keeping them indoors more. But this doesn't always work, and it's hard to prevent a determined cat from getting out. Collars with bells may help, but efficient hunters can still catch prey (especially young birds). If you must let your pet cats go outdoors, it's the best to do it in daytime when rodents are least active and to keep them inside during the time of year (usually spring and early summer) when young nestlings are fledging and are still inefficient flyers. And of course for cats that go outside spaying and neutering are crucial to keep their populations under control.

The next time you notice your innocent-looking kitty purring on the sofa, remind yourself that this predator just *looks* like a couch potato.

Émilie Lantin

CAFE CONVERSATION

Discuss some cat-friendly strategies you could encourage to help control cat predation on birds and small mammals in your neighborhood. Do some laws already exist?

13

PIGEONTOWN

From Times Square to the Champs-Élysées, from Oxford Circus to the canals of Venice, they squat on abandoned buildings and ledges, in parks and subway entrances, omnipresent in their street-colored suits. Some people give them handouts, while others see them as winged rats that should be eradicated. They're found in all our urban ecosystems; sometimes a few, but often too many. Exploring what governs pigeons' presence across neighborhoods within a city reflects one of the fundamental preoccupations of the science of ecology: the distribution and abundance of living things.

Living organisms are not uniformly distributed across our planet. You can see that grass blades form clumps within your lawn and that lawns themselves are separated by expanses without grass. The distribution of animals is just as nonuniform. But, unlike plants, they can move around, and their distribution depends on the choices we call behavior. **Behavioral ecology**, then, is a basic and important component of the study of animal distribution and abundance.

Pigeons are mostly found in urban settings and are absent from most other major ecosystems, but they are not uniformly distributed

in a city. You may encounter fifty in one square, perhaps twenty on a university campus, a hundred or more at a downtown subway station, and maybe thousands outside an industrial flour mill. What factors explain such patchy distributions? Do the same factors govern the spatial distribution of fish off the coast of eastern North America, grizzly bears in the Rocky Mountains, or snow geese in the High Arctic?

WHO ARE THESE URBAN PIGEONS?

The pigeons that hang around all year long, tirelessly reading inscriptions on monuments, all originated in the Mediterranean basin. They were domesticated during antiquity, mostly as food for the nobility, until people discovered chickens were tastier. Pigeons were also used as messengers. Before the invention of the Internet (or the phone, for that matter), people traveling a long way carried along a few pigeons from a correspondent's roost and sent back messages by airmail. This worked because the "homing" pigeon's goal was simply to get home as quickly as possible. Pigeons were also used for racing, and people often gambled substantial sums on them. Finally, because pigeons could be crossbred to develop extravagant colors and shapes, they were also kept to please hobbyists (Charles Darwin himself was a pigeon fancier). Today the pigeons that loiter on dirty city streets are the descendants of those very birds that chickens replaced on dinner plates, that lost too many races, or that were not ornamental enough.

COSTS AND BENEFITS OF GROUP LIVING

One rarely sees a pigeon alone. Given that they are **monogamous**, one would expect to encounter pairs. But most pigeons are seen in small groups because they are also **gregarious** and breed in colonies. Some scientists point to the evolutionary advantages of group living. For example, it is easier to detect danger and avoid being a predator's victim. The beautifully synchronized flight of a flock of birds circling in the air demonstrates the challenge to a predator that tries to catch one. You'd have the same problem catching a ball if a dozen were thrown to you all at once. But while these evolutionary advantages can explain why pigeons live in groups, they still don't explain why we find a hundred together in one place but only twenty in another. But the cause of variation in group size lies not in the benefits of group living but in the costs.

Most costs of group living result from increased competition for limited resources (food, nest sites, mating partners). The more individuals living in one location, the smaller each one's share. Variation in the resources in different places could account for changes in group size. Pigeons should thus choose to minimize competition, ideally by adjusting group size based on the food (or other resource)

available. This principle underlies the theory of the **ideal free distribution** developed in the mid-1970s to explain animal group sizes. Let's explore this idea further.

ARE PIGEONS IDEAL AND FREE?

A simple experiment conducted with a friend or two can address whether pigeons are ideal and free. Prepare some food that is easy to hand out and monitor: a loaf of stale bread broken into pieces will do perfectly. Find a pigeon group that is small enough to count easily (perhaps thirty or fewer). Attract the pigeons by throwing a few crumbs on the ground; they're usually hungry early in the morning or toward the end of the afternoon. If pigeons come when you throw some crumbs, you're ready to begin.

Split the bread equally with your partner and stand a few yards apart. It's useful to have a third person who can stand back and record the number of pigeons each of you attracts. Each person should throw small pieces of bread at that same agreed-on rate, for example, every five seconds. You and you partner now represent two "habitats" the pigeons can choose between.

If your pigeons are ideal and free, they should distribute themselves according to the food available at each location. Note the number of pigeons in each habitat at regular intervals, say every thirty seconds. Ideal and free pigeons will distribute themselves according to the rate at which you and your partner are providing food. If you throw food at the same rate, the pigeons should divide into two equal groups. If at some point your partner starts throwing food twice as fast as you do, the distribution should change until there are twice as many birds in that habitat. You could then reverse the habitat profitability and see how long it takes the pigeons to reach a new stable distribution.

This experiment has been conducted many times with students in behavioral ecology classes. Each time, it is impressive to see to what extent the pigeons manage to distribute according to the predictions of this theory.

FROM THE PARK TO THE CITY

Pigeons can be "ideal" when two people hand out food a few yards apart in a park. But are they "ideal" at a much larger scale, such as across neighborhoods? Do they know the quality of the alternative food sources? Of course, pigeons living downtown won't know how much food is available on the other side of the city. But research in Montreal has shown that each pigeon belongs to several flocks. An individual may spend the morning with one flock, the afternoon with another, and join a third in the evening. Each one has an idiosyncratic itinerary taking it from one flock to another. Without knowing about all the sites in a city, each pigeon clearly tries out many different ones, becoming aware of the resources at each locality.

Pigeons are also "free" to move between flocks. Yes, you sometimes see them fighting, but these fights are not about group membership. Pigeons do not gang up to evict a group member or keep another from joining. Thus, pigeons appear ideal and free at a local scale, so their distribution throughout the city reflects rather precisely the daily quantity of resources available in each locale.

FROM PIGEONS TO HIGHWAYS

We can also apply the theory of ideal free distribution to birds visiting neighborhood feeders. If you increased the quantity or quality of food at your feeder, how would this affect the overall bird distribution? Should it affect the total food each bird eats? The theory predicts that new foragers should show up at a more richly provisioned feeder. But ecological studies show that each of the visitors at your deluxe feeder will get the same amount as each of the smaller number visiting your neighbor's more frugal one.

This way of thinking about distribution can be applied to many urban inhabitants. One case to consider is the distribution of cars on urban highways, which depends mostly on the flow capacity of a particular road. Would it take less time to go between the city center and the suburbs if all drivers were ideal and free—that is, informed? Could this be why electronic signs and smartphone apps giving up-to-date information about traffic patterns (habitat quality) are becoming more common? The ideal free distribution can help us choose which line to join at the supermarket or decide whether to change lanes in heavy traffic. The more you consider it, the more applications you'll find for this ecological theory. Don't forget, however, that the same principle should apply to the distribution of all animals—as long as they are ideal and free.

Luc-Alain Giraldeau

CAFE CONVERSATION

Do you think the theory of ideal free distribution applies beyond pigeons? Could you use that theory to predict, say, whether people who change lanes in heavy traffic really get where they're going faster? Using the logic of the ideal free distribution, what would likely happen if you widened some street or highway to eliminate its chronic traffic jams?

IDEAL FREE DISTRIBUTION

This theory simplifies the world to an extreme, as is often done to clarify a point in science (consider how much physicists have learned from the perfect pendulum, which unrealistically has neither friction nor resistance). Such simplifications do not mean scientists think such perfect conditions really exist, only that is simpler to understand the principles of a perfect system before explaining the workings of a real one.

Ecologists use the same approach. In the case of animal distributions they imagined individuals that were *ideal* (possessing perfect and complete information about the quality and whereabouts of all available habitats) and *free* (able to move to any of these habitats without cost). If we allow ideal and free individuals to distribute over habitats, assuming each wants to maximize individual gain, we end up with a distribution where no one can move to another habitat without doing worse. Individuals will then be distributed according to the ideal free distribution: the number in each habitat will be proportional to the local habitat's richness—an ideal world.

14

ANIMAL INTELLIGENTSIA

Two pigeons sit on a window ledge at the art museum, having a private discussion apart from the main group.

"I looooooove that Picasso!" the one on the left exclaims enthusiastically.

"You are so wrong, my friend. That's a Monet," replies the other with disdain.

"I'm sorry, but that painting is definitely a work of Picasso," insists the first.

Hard to believe in such a conversation? Well, it shouldn't be. Behavioral ecology researchers in Japan have demonstrated that pigeons can tell the difference between a Picasso and a Monet. Their visual acuity and their mental aptitude allow them to perceive differences in painters' styles, memorize the differences, and categorize them mentally. They can identify the work of an artist (by pushing a button, for example) even if they have never seen that specific painting before.

Pigeons don't spend a lot of time on art appreciation, however.

They use their ability to visually categorize information mainly to distinguish between types of food, birds they meet, and other common encounters in their daily life. Over time the specific characteristics of pigeons' environment have without doubt contributed to these cognitive abilities. In fact, all animals can be said to have a full mental life, allowing them to react to their environments. But because they cannot tell us directly, we have to carefully watch for clues to such intelligence by conducting experiments.

THE BRAINY ANIMAL

Does the concept of intelligence apply to animals? And if so, does it exist for them all or just some? How and why could this intelligence have evolved? To answer such questions, first consider what we mean by intelligence. The *Oxford English Dictionary* defines it as "the ability to acquire and apply knowledge and skills."

It's not really a stretch to realize that many animals can meet this definition. Stories abound of lost cats who found their way home over extremely long distances. If you have a bird feeder in your yard, you've undoubtedly noticed squirrels' innovative ability to steal seeds from almost any "squirrel-proof" feeder.

WE THINK WE CAN

We need to interpret such examples with caution, however. Humans tend to attribute human feelings, emotions, and intentions to nonhumans: these attributions are called **anthropomorphism**. If we are to remain objective, the scientific study of animal life and behavior must remain dissociated from the human point of view. It can also be difficult to establish a single "intelligence scale" for all animals, allowing us to compare sea lions with sheep, or bears with ants. So how do behavioral ecologists measure animal intelligence?

To start, we must distinguish between the various mental processes (**cognition**) that make up intelligence and evaluate them separately. For example, the renowned intelligence quotient (IQ) for humans tries to rate our abilities to memorize, to understand concepts, and to organize spatial relations. To evaluate such cognitive abilities among animals, we need a separate test for almost every species because they are so different. Instead of giving a written or oral exam, researchers must ask a precise question and try to create a situation in the laboratory or in the wild that will let the animal answer by its behavior. Such tests can tell us about an animal's range of cognition within a given context. Here are some examples of questions and their animal answers drawn from situations you may see around you in the city.

How Does a Hummingbird Find the Best Feeders?

Does it use cues like the color and shape of the feeder or its position relative to a house or to other food sources? Hummingbirds do, in fact, process, perceive, and memorize color, shape, and position. But not all information has the same importance: position is the primary cue. If you move a feeder, hummingbirds will return to the original place before searching elsewhere. Next the bird will use physical characteristics to identify a feeder, starting with its shape and finally its color. So you don't need to add red food coloring to your hummingbird food. Not only is it unhealthful for the birds, it is one of the least important cues they need.

Because hummingbirds naturally feed on nectar from stationary flowers, using physical placement of the food source as the primary visual clue makes sense. The position of a patch of flowers the hummingbird encounters in the wild will vary less than the color and shape of the flowers.

Are Insects Intelligent?

Because insects are so small, we don't often think of them as having intelligence. Don't be so quick to judge! Bumblebees, for example, are incredible visual decoders. Based solely on shape, they can learn to distinguish flowers that still contain nectar from those that are empty. They can even distinguish when they see only part of a blossom or when they view it from a different vantage point.

Such flexibility helps them search for food more efficiently. It likely evolved to let bees optimize nectar collection in a diverse environment even though, like all insects, they have only a cerebral ganglion instead of a real brain. With a cerebral ganglion no bigger than a pinhead (weighing about 1 mg), a bee can use minuscule details of flowers to quickly identify which will give it the most food. An amazing feat.

Does My Dog Have Good Short-Term Memory?

Here's a memory test to try on your dog. Give her a toy (a bone, a ball, a rubber chicken—whatever you like). Then take it away and let her see you hide it behind or under one of four boxes or similar objects. Then put her behind a barrier (a screen, a wall) that hides all the boxes and make her wait 0, 30, 60, 120, or 240 seconds in several trials before letting her to go search for the toy. You should find that even after several minutes the dog will remember relatively well where the object was hidden. That she still searches even after some delay demonstrates that she has maintained a memory or neural representation of the toy—**object permanence**.

In human children the ability to remember hidden objects doesn't appear until age two. For dogs and other canids (wolves, coyotes, foxes, and jackals), this ability is crucial for successful hunting because prey often disappear or hide behind obstacles like a bush or a hill. Even if our domestic dog is more likely to hunt tennis balls than her next meal, she has inherited this ability from her wild ancestors.

GRAY MATTER ACCORDING TO DARWIN

The intelligence of animals allows them to get around and to use their environment more efficiently. Information is used to find better food and shelter as well as to raise young successfully. As you might suspect, however, there can be large differences in the mental capacity of different species as well as between subspecies. Such differences help us understand the functioning and evolution of such mental faculties. Each animal, like each person, is unique, the result of a perpetual interaction between the genes it inherits from its parents and the environment where these genes express themselves.

How did animal species develop such different capacities? Scientists believe that not only are living organisms' physical attributes under pressure from environmental constraints, but so are their mental capabilities, which are the basis for their behaviors. It follows therefore that the ecological conditions an animal experiences will shape its sensory and cognitive abilities.

Toward such evolutionary understanding of animal intelligence, many ecologists study cognition by examining how animals use their senses to gather information from the environment, how their central nervous systems (the brain) treat such information, and how they modify their behavior—an interesting intellectual challenge for humans.

France Landry

CAFE CONVERSATION

Everything seems to be smart these days: smartphones, smart cars, smart homes. Is there any difference between a smart telephone and a smart cat? If so, what is it?

15

SQUIRRELLY NEIGHBORS

Wandering through a forest, you are very likely to hear an angry squirrel chasing an intruder from its territory. Whether that's some other squirrel or an oblivious human, an angry squirrel sounds impressive. There's no doubt that squirrels guard their territories with enthusiasm and vigor.

Animals protect territories used for food, mating, or refuge (e.g., nest sites). Hummingbirds establish food territories during migration that give them exclusive access to a supply of nectar. Gulls defend tiny reproductive territories in the midst of large colonies. There is no food there, just enough space to build a nest that a gull protects ferociously against often cannibalistic neighbors. Finally, several species defend territories both to protect their nests and to

guard food resources, like the gray squirrel that is the focus of this chapter.

Such all-purpose territories give the animals who hold them all the essentials to ensure not only their own survival, but that of their descendants. The first requirement is shelter, so squirrels' territories must contain places to build nests of leaves and twigs called dreys—in the forks of high tree branches, in natural tree cavities, or in holes left by woodpeckers.

A squirrel's territory must also contain food. Trees are the primary source, providing nuts, seeds, seedlings, and sometimes other fruits. Although they will occasionally snack on insects, eggs, baby birds, and small amphibians, squirrels are essentially **granivorous** (feeding on seeds).

Finally, a squirrel has to reproduce, as all animals are eventually driven to do, so it requires access to mates. Some species also must find a suitable place to raise young; the female tree squirrel will use the nest she has already built for shelter.

Resource density, the quantity of resources available in a given space, will affect the size of an all-purpose territory like the squirrel's. The usable or realized density of resources is determined by their quality. For example, nuts are high-energy food but are usually rare and less densely distributed throughout the territory than smaller seeds of lower quality. As long as the squirrel can be meet its needs within its territory, it will not have to leave it.

In dense forests where trees are close together, territories will be smaller because squirrels can find all the food they need in a smaller area. Because moving around takes energy, limiting the distance traveled while foraging optimizes energy intake. Another benefit is that traveling less lowers the risk of running into a hungry predator. But too much prudence can be risky too: a squirrel that never left its nest would be unlikely to meet its nutritional needs, making it also vulnerable to predators.

MASTER OF THE HOUSE

In some cases the boundaries of a territory are defined primarily by interactions with neighbors of the same species and result from competitive pressure or from conflicts between new arrivals and individuals already settled there. The intensity of such interactions defines how territorial a species is.

When neighbors live far apart so territories are not contiguous or overlapping, a species is likely to be tolerant. The cost of maintaining the territory is minimal, and the animal can easily adjust its size. Neighbors whose territories are contiguous but do not overlap will ferociously defend them against competitors. Even when there are more resources than it needs, a territorial individual will protect the larger space, apparently as a safeguard against times when they become less abundant.

Competition between neighbors is linked to resource density. When animal populations are large, each individual's territory is reduced and becomes the object of conflict. For this situation to be viable, resources have to be abundant enough. When they are scarce, the population density will also be low, and it is less essential to defend territories. It's a question of balancing the energy invested in defense against the payoff from extra resources.

AN UNNEIGHBORLY SEASON

For most of the year, gray squirrels are normally not extremely territorial; they tolerate each other as long as the neighbors do not trespass too often. The one part of their territory that they particularly guard is the center or **core**. The edges of squirrels' territories often overlap with little conflict.

The situation is completely different during the reproductive season, toward the end of winter and into spring, when competition within the same sex intensifies. Females are not receptive to males for long, nor are they all receptive at the same time. Males will compete for receptive females, increasing the chances of an aggressive encounter, with the strongest rivals holding the others at bay to keep females to themselves. These dominant males guard the largest mating territories possible to have access to more females.

The mothers-to-be also compete to have their offspring sired by males from the best territories, where food density is greatest. A population's reproductive rate is the average number of young born to each female and surviving to adulthood. This rate is highest when food is plentiful, because a well-fed mother can give birth to more and healthier babies. Choosing the proper territory thus is crucial to a squirrel family's long-term survival.

AN URBAN TRUCE

In the city, the concept of an exclusive territory is vaguer. Urban squirrels needn't worry too much about predators. Moreover, cities are full of gastronomic delights. Trees in parks provide seeds, just as they do in the forest, but garbage cans overflow with succulent morsels. And don't forget all those people handing out peanuts and candy, seduced by those cute beady eyes. These extra sources of food disrupt squirrels' normal behavior, so that several unrelated individuals will often sit together in the same tree—which would rarely, if ever, occur in the forest.

Squirrels' territorial and competitive behavior can therefore be almost completely suppressed where accessibility of resources is not a problem. This is yet more evidence that territoriality is not a fixed feature but a plastic behavior based on the costs and benefits of maintaining exclusive access to a parcel of land.

Simon Bilodeau Gauthier

CAFE CONVERSATION

Do humans have territories? If so, what type are they? Is there more than one type?

16

FROM THE CHIMNEY TOPS

Chimney swifts occupy a unique ecological niche, and a very urban one. In the past this species nested in dead hollow trees, typical of precolonial forests, but now it depends more on chimneys for nest sites. Ironically, despite their ability to adapt to these human-created habitats, these birds are becoming rarer.

The chimney swift (*Chaetura pelagica*) belongs to the family Apodidae, a group that has evolved body shapes and flying techniques similar to those of swallows, though they are not related. In fact, swifts are closer to hummingbirds than to swallows. There are about a hundred swift species throughout the world, distinguished by long, pointed wings that arch backward and by an aerodynamic body shape, making them expert flyers. They are exclusively airborne hunters, spending most of the day catching insects high in the sky. Despite these amazing wings, their feet are not well developed,

and they have tiny toes, earning the family's Latin name Apodidae, "without feet."

Several specific adaptations make the chimney swift perfect for literally hanging out on chimneys walls. Like bats', their sharp, curved claws let them grasp rough walls. The longest tail feathers have a rigid central shaft or rachis that supports the bird on a vertical surface—a characteristic woodpeckers share.

Chimney swifts build their nests from twigs attached directly to walls, using glue made from saliva. They share this sticky saliva with the white-nest swiftlet of China, whose nests are harvested in large numbers to make bird's nest soup (often incorrectly called gull's nest soup).

Despite its predisposition for city life, the chimney swift was once able to meet all its ecological needs in the forest. Before Europeans settled in North America, swifts sheltered in a variety of natural vertical structures: rocky crevices and caves, but mainly hollow trees. The rotting out of large tree trunks creates natural chimneys that ecologists believe these birds used for nesting and sleeping. Because modern foresters often cut down every dead tree, large hollow trunks are now rare, forcing these swifts to use substitutes like silos and other farm buildings—and, of course, chimneys.

A DEMANDING RESIDENT

Despite the apparent abundance of chimneys in city and country, only a few meet all of swifts' requirements. First, so as not to be burned alive, they must find chimneys that are unused. But because they are also sensitive to cold, they prefer ones that remain open at the base, so warm air rises from the building below. Furthermore, to avoid getting stuck inside, swifts will use only chimneys wider than 12 inches (30 cm). And they prefer sites not too far from water—a stream, pond, or lake—so they can take advantage of the plentiful insects that emerge there.

An entire population of swifts will require many chimneys. The biggest will be used communally as dormitories when swifts arrive in the spring, and again in the fall once their young have fledged. Sometimes the dormitory chimney will also be used for nesting, but

generally by only one couple—the rest have to find others. A pair will use the same chimney year after year, so the number available is an important constraint limiting population size.

THE CHIMNEY SWEEP STRIKES AGAIN

Before electricity became common in homes, most chimneys were used both for heating in the winter and for cooking throughout the year. Only those used solely for winter heating were available to swifts during the nesting season. Paradoxically, the very increase in human settlement that led to the gradual decline of their natural forest habitat probably saved these swifts from extinction by providing plenty of replacement nest sites.

Over the past few decades, however, those many chimneys that used to stick up on the urban horizon have been slowly disappearing, leading to a precipitous reduction in the number of chimney swifts. This is an interesting example of a species decline owing to changes not in natural habitats but in artificial structures.

Many older chimneys have been blocked up or torn down, either in renovation or to reduce maintenance costs. Also, in many municipalities homeowners must install a device to keep sparks from flying out of their chimneys, or their chimney sweeps may advise them to install wire mesh. Both will keep the birds out.

Many new residences have no chimneys, or they are so small that they become deadly traps for birds that try to use them. Furthermore, many of the biggest chimneys—those of schools, hospitals, churches, and other institutions—are being converted to accommodate natural gas heating. This requires a metal liner that is too smooth for swifts to grasp or attach their nests to.

Another important factor is that sweeping chimneys, traditionally done in the autumn, now occurs year round, destroying nests and killing young birds.

THE LIGHT AT THE END OF THE CHIMNEY?

Like urban species, the chimney swift has without doubt benefited from human-built structures. But in contrast to many others that are generalists and have shown great adaptability, its success has been strongly linked to a specific artificial structure that was once very common but is becoming rare. This dependence on chimneys came about because of the removal of their natural nesting sites. Now, as chimneys also disappear, the species could be saved by newer forestry practices, such as ecosystem management, that promote leaving dead hollow trees standing. Until this happens on a large scale, however, the immediate future of the chimney swift will depend mainly on changes in the availability of chimneys in urban and rural environments.

For now old neighborhoods in towns and cities, as well as public buildings like unrenovated churches, have become important refuges for chimney swifts. Swift populations should benefit from making homeowners aware that summertime chimney sweeping and renovation that destroys chimneys endanger the birds. In many places in North America, observatories have been set up at important swift dormitory and nesting sites. These programs not only educate the public but preserve chimney swift habitats. If you visit one of these sites, observe the beauty and agility of these little birds' flight as they come home at the end of the day.

Antoine Nappi

CAFE CONVERSATION

What might be some ecological or scientific justifications for preserving chimney swift habitats? How about habitats of other organisms that might be endangered?

17

LONG LIFELINES

Eastern North America, 1640s: temperate deciduous forests cover the landscape seen by most early European explorers. Slowly, new buildings replace the forests, the start of an expansion that will lead inexorably to a decline in the trees being cut to build and heat houses, to lay walkways, and to construct boats for sailing back to Europe. Islands of trees that were never cut down are a mine of information on what those original forests might have looked like. Trees in the city can likewise tell us much about their history and ours.

In seasonal environments, trees grow only when it's warmer and enter a dormant state when it's cold. This alternation of growth and dormancy produces growth rings—one per year. Growth rings arise from the cambium, a thin layer of living cells under the bark that are activated each spring and make new wood cells. The cells produced at the beginning of summer are large and pale, with thin walls, serving principally to transport water up the trunk. Toward the end of the summer, the cambium produces darker, smaller cells with thicker walls as structural support for the growing tree.

A complete tree ring therefore comprises a pale zone and a dark zone. A tree trunk grows in circumference every year by accumulating these rings just under the bark, easily seen with the naked eye on freshly cut wood. Only the cells produced in the past few years are alive; most of the trunk consists of dead cells.

Happily, we can look at tree rings without cutting down the tree. Ecologists who want to date a tree use a Pressler auger also called an increment borer) to extract a small core of wood 0.2 inch (5 mm) in diameter along the radius of a living tree, then count the alternation of dark and light cells on this core to determine the tree's age—and much more. By studying growth rings, ecologists can date particular events that might have happened directly to the tree or in its immediate surroundings, reconstructing past disturbances and climate. This is the basis of the branch of ecology called **dendrochronology**, from the Greek *dendron*, "tree," and *chronos*, "time."

FROM WOOD COMES MEMORIES

The growth of a tree is affected by weather. A bad year will produce a thin ring; good years produce thicker ones. But climate is not the only factor affecting tree growth. Outbreaks of insect pests that eat all a tree's leaves will also result in a thin tree ring for the year, or none at all. Freezing rain, fire, flooding, or a light gap opening in the forest canopy will all be recorded in the rings. And all the area's trees of the same species will produce the same growth pattern.

Tree ring width can vary substantially between species and from site to site. For example, a silver maple (*Acer saccharinum*) growing in the rich, wet environment of a floodplain produces an annual ring 0.39 to 0.47 inch (10–12 mm) wide. Eastern white cedars (*Thuja occidentalis*) that commonly grow on the sides of cliffs where soil is very poor and dry produce rings of, at most, 0.04 inch (1 mm) each year.

Large-scale environmental events, like the warm years of El Niño, affect tree ring growth, but so do more local events. And this is true whether a tree grows "naturally" in a forest or has been planted next to a city sidewalk. Events that might seem inconsequential to us can have important effects on trees. A pile of earth that deprives tree roots of oxygen, or a snowplow or springtime freezing rain that damages the trunk or branches, can make a tree use up its energy resources and reduce growth.

WHEN THE WEATHER IS LATE

Occasionally ecologists will find an ancient stand of trees that they can sample to reconstruct past climates. From such finds, dendrochronologists can sometimes get detailed information about the timing of conditions by comparing the growth of tree rings with long-term weather records. For example, some species including the sugar maple (*Acer saccharum*) and Douglas fir (*Pseudotsuga menziesii*) grow fatter rings when there is a lot of rain in early summer. Thinner rings result when the previous July was hot. When these trees are exposed to a long hot period toward the end of summer, while they are accumulating resources for the future, it will affect the growth ring for that same year.

In this way ecologists can reconstruct the probable climate that people living in much of North America experienced over several hundred years. In general, winters were colder, especially in the middle of nineteenth century, during the Little Ice Age. This colder period has been detected in several dendrochronology studies.

In most North American cities the oldest trees disappeared long ago. A creative solution has been to sample wooden beams from old houses, or logs used to build cabins. **Dendroarchaeology** essentially dates wood from human buildings to reconstruct past civili-

zations. By matching the tree ring record with historical texts, scientists can infer climatic conditions in particular places. Trees thus form the basis of record keeping: not only do they produce the paper we once wrote our history on, but their rings preserve the history of the environment.

Danielle Charron and Yves Bergeron

Bamboo is used more and more in our houses for everything from flooring to cutting boards and cooking utensils. Look around your house (or the local kitchen shop) and compare the grain patterns on a cutting board made of bamboo and one made of more traditional wood, which differ because bamboo is a grass and belongs to a completely different plant group (the **monocots**) than trees. Traditional wood comes from **dicots** and has true rings, while bamboo stems are hollow and have no rings other than the outermost one. You can learn more about bamboo, including how it is made into products we use and why the grain looks the way it does, by searching on the Internet.

THE THOUSAND-YEAR-OLD CLUB

The oldest known tree in North America grows in California and been named Methuselah. It is an ancient bristlecone pine (*Pinus longaeva*) that dendrochronology has shown to be about five thousand years old. The original seed of this gymnosperm would have sprouted before the Egyptian pyramids were built, and even before the extinction of the mammoths. The oldest trees in eastern North America are the eastern white cedars (*Thuja occidentalis*). But older does not necessarily mean bigger: these ancient cedars are only about 12 inches (30 cm) in diameter. In western North America, older trees are often yellow spruce (*Chamaecyparis nootkatensis*) or Douglas fir (*Pseudotsuga menziesii*).

PART III
The Metropolis

18

CANARY IN THE CITY

Lichens form colorful crusts on old tombstones and large round tufts on northern soils. They come in a variety of colors and shapes. Aside from their beauty, we're going to explore what they can reveal about the health of the environment we live in. As long as the air is pure, lichens can inhabit environments from the ordinary to the bizarre, and they tell us a lot about a city.

Lichens, often confused with mosses, are the result of a **symbiosis** between algae and fungi. The Greek term literally means "living together." In lichens, algae cells are wrapped in a fungus that protects them. Whereas mosses usually feel soft, lichens are often crusty to the touch and come in a variety of colors besides green.

What first brought ordinary fungi and algae into a symbiotic relationship is an evolutionary mystery; lichenization is not yet completely understood. What ecologists do know is that not only can several species of unicellular algae be part of the same lichen, but a single algal species can be found in several species of lichen. On the other hand, the associated species of fungus allows us to identify a

lichen. Generally the fungi that form lichens are very small species belonging to the phylum Ascomycota. Morels and truffles are also members of this very large taxonomic group in which all species ripen their spores inside small sacs called **asci**.

LIVING TOGETHER

The symbiosis that links the two species within a lichen is mutually beneficial and is therefore a mutualism. The alga supplies energy to the fungus through photosynthesis. All plants and algae use photosynthesis to convert sunlight, carbon dioxide, and water into sugars and oxygen (see also chapters 1 and 10). The fungus, in return, protects the algal cell and keeps it in a position where it can obtain sunlight. The fungus forms filaments (hyphae) that tightly grip the substrate (rock, soil, or bark) the lichen grows on.

Furthermore, the fungal partner can form chemical compounds that benefit the lichen. Lichens produce over three hundred substances that serve roles such as inhibiting overly fast algal reproduction, filtering light radiation, and protecting the partners through antibiotic properties or by repelling herbivores. Humans use these compounds in medicines and cosmetics, specifically antibiotics, perfumes, and paints.

In contrast to a mutualism, an ecological association or symbiosis that benefits only one partner is called **parasitism**. Many microscopic organisms that make us sick, like viruses, bacteria, and fungi, show this type of ecological interaction, but several of these small organisms can also interact positively with animal species. Cattle and many other herbivores, for example, could not digest the cellulose in plants without certain bacteria and other unicellular species like protozoans. And some microscopic fungi attach to the roots of plants to form mycorrhizae that help the plant absorb water and nutrients. The fungus in such a mutualism is rewarded with the sugars and vitamins the plant produces.

Symbiosis that benefits only one partner and does not really affect the other is **commensalism**. An example is a bird that builds its nest in a tree: beneficial to the bird, no effect on the tree. And lichens? Even though a lichen results from a mutualism, its growth on the tree's bark is mostly a form of commensalism: good for the lichen, little effect on the tree.

A SUPERFICIAL SENSITIVITY

To assess the air quality in your city, look at the bark of old trees. Depending on your neighborhood, there's is a good chance some lichens are growing. Lucky you! Lichens are so sensitive to damage from pollution that they are used as **bioindicators** of air quality in many cities around the world.

Lichens live essentially from the light and water they obtain directly from their environment. Because they have no way to regulate or filter these resources, they directly absorb all particles suspended in the air and in rain, including all pollutants. Even far from cities, lichens can be affected. In 1986, thousands of Lapland reindeer had to be killed because they ate lichens that had absorbed enormous quantities of radioactive isotopes after the meltdown of the Russian nuclear reactor in Chernobyl, thousands of miles away.

Ecologists measure the accumulation of various toxic compounds like lead, fluoride, and sulfur dioxide in the **thallus** (the fleshy body) of a lichen to determine the degree of pollution in the air. Over the long term, these pollutants will kill many species of lichen, and ecologists can map pollution zones by noting the presence or absence of lichens in particular areas. In larger cities the most industrialized areas have the fewest lichens, but sometimes residential ones have the fewest, while large city parks will harbor many species.

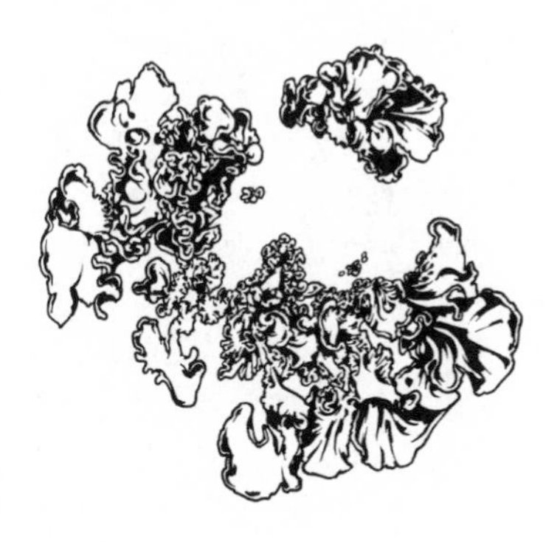

Although more sophisticated methods now exist to sample air pollution, lichens provide a simple, cost-effective way of getting similar information over a longer period. Furthermore, by mapping them we get an overview of the geographic distribution of pollution. How healthy are the lichens in your city?

Héloise Rheault

If you or a friend plans to move to a new neighborhood in the same city or a different one, you might want to consider health in your search for a new home in addition to price and safety. Take a walk around potential neighborhoods (or your own, if you aren't moving). Census the lichens on the trees you pass. What proportion of the main trunk of each tree has lichen cover? What proportion of trees in the neighborhood have lichens at all?

INDICATOR SPECIES: HELPING ECOLOGISTS

Some species are more useful than others for detecting environmental changes. It would be a huge undertaking to inventory all the species in a particular location to assess its ecological health, so ecologists often use **indicator species** to develop conservation plans:

Bioindicator: These species respond rapidly to disturbance that will also negatively affect most other species in an area. Lichens and amphibians are usually the fastest to respond to air pollution and water pollution, respectively.

Keystone species: The presence of these species is crucial for the survival of most other species in a habitat. Beavers construct dams that create new habitats essential for many other species.

Umbrella species: These species require large habitats for food, lodging, and reproduction. By protecting the habitat necessary for these targets, ecologists can assure that other species with lesser habitat requirements will be saved too. Woodland caribou need large expanses of boreal forest, which is also the habitat of many other animals and plants.

Rare species: Sometimes it is necessary to conserve an exceptional region because it is the habitat of a rare or endangered species. Some fish species that used to be widespread are now found only in a few river basins, necessitating the conservation of those rivers, like the Fraser River and other Pacific Northwest sockeye salmon runs.

Flagship species: These species have a particularly high economic or social value. It is sometimes easier to persuade people and government agencies to conserve a region if it contains a popular species. The health of whale populations is more likely to elicit a strong response than is danger to an insect population—or lichens, for that matter.

19

WING BEATS

It's early spring, and each day gets noticeably milder. A chrysalis hangs under a dead leaf, attached by delicate strands of silk. Suddenly it splits open, and a butterfly struggles desperately to escape. First she sticks out her head and antennae. Then, using her feet, she extracts her abdomen and finally her wings. She prepares to explore the city anew, as she couldn't do when a caterpillar.

During the long sleep of a metamorphosing insect like a butterfly, its chrysalis is well camouflaged, and only a sharp-eyed bird is likely to distinguish it from a tree branch. But as it emerges, the insect is at the mercy of predators, unable to escape. Not yet a butterfly that can fly away, it looks like a crumpled ball of crepe paper. Its new wings will need to dry out before they can be used. After about an hour, an eternity for a butterfly, it can finally fly away.

The butterfly whose story makes up this chapter is a female with a wingspan of about 1.5 inches (42 mm). Her wings are white with three black spots on the top. She is a member of a very common species that you have surely seen in the city: the cabbage white butterfly (*Pieris rapae*).

An adult butterfly can, in theory, fly more than half a mile a day, but we're unlikely to ever find her so far from her starting point. She will doubtless have been forced to change her route—turn left, turn right, and so on—because of obstacles in her path. Flying straight from point A to point B is not often an option for an urban butterfly. When ecologists try to predict how far an animal will go in a day, they must consider how easily it can cross a given piece of land. The cabbage white flies in all sorts of habitats, but it will take longer to fly through a forest of buildings than over an open field.

Our new butterfly leaves to explore her environment. In open

spaces like parks, gardens, and vacant lots she can fly long distances; otherwise she will have to get around houses and buildings by using streets, while avoiding cars and people. For her the urban landscape is fragmented. Islands of green are more or less isolated from each other but sometimes connected by green corridors like trees planted along the sidewalk or by vegetation alongside a highway or railroad tracks.

CABBAGE PATCH BUTTERFLIES

Like humans, butterflies' principal sense is vision (as opposed to moths, which are out at night). So our cabbage white depends on light to accomplish all her activities, mainly collecting nectar and reproducing. Although she does not have acute eyesight, her globular eyes give her a large field of vision. She has other helpful senses, too; she uses her antennae to touch and smell and to detect vibrations and air currents, information she needs for effective navigation and to keep her balance and orient herself while flying. The cabbage white also uses sensitive hairs on her feet to find and "taste" nectar.

Butterflies' wings allow them to search for food efficiently and, also important, help their offspring disperse. Females disperse farther than males, which stay close to where they were born and thus have the opportunity to mate several times. Females, which are so

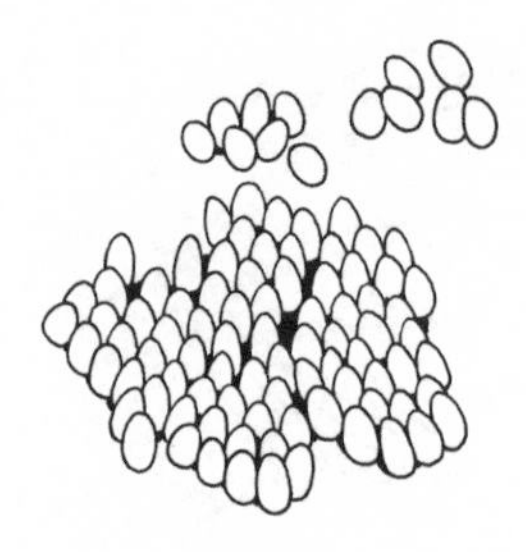

busy flying around, usually have only one chance to mate. Mating is rather quick, with no fancy courting rituals or nuptials. After mating, the female flies off to find a site to lay her eggs.

Because she flies around so actively, the female can lay her eggs across a large region, but only where there are enough host plants, mainly the cabbage family (the crucifers). Cabbage whites prefer to lay their eggs in several places to avoid having all the offspring die if something goes wrong. Our butterfly knows better than to put all her eggs in one basket.

FUTURE CABBAGE WORMS

The life expectancy of a cabbage white is five to fifteen days, just long enough to ensure mating. Our female will find several places to lay her eggs: nasturtiums growing in window boxes, cabbage leaves in a kitchen garden, and perhaps in a vacant lot, but only about 5 percent to 15 percent will hatch and survive to adulthood.

In this story our butterfly was lucky. Almost all the eggs she laid on the nasturtiums and in the vacant lot hatched. Sadly for her, those she laid in the kitchen garden did not survive. The gardener, eager for beautiful cabbages, handily removed all the voracious caterpillars that hatched.

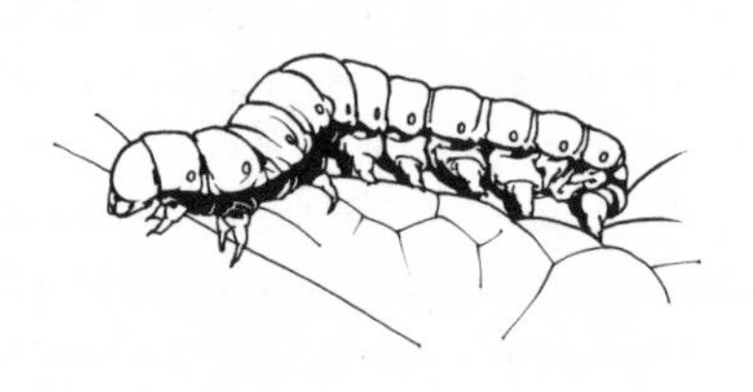

Forty-five days after being laid, our butterfly's surviving children will fly through the streets searching for nectar and laying sites. In August or September the eggs of this second generation will hatch into hungry caterpillars that form new chrysalides. But this time no new adults will emerge right away. Because the long winter is near, these chrysalides will enter a dormant phase called the winter **diapause**. The third generation of butterflies will not discover the city's streets, houses, and gardens until early next spring.

Samuel Pinna and Nathalie Roullé

Look at a satellite view of your city on the Web. Do you see places where the greenery forms corridors? Or are your city's parks relatively isolated? Can you see places where corridors between green patches might be enhanced?

LANDSCAPE ECOLOGY

The story of the cabbage white butterfly is an excellent demonstration of how important spatial structure is in the ecology of species, especially those we live with in cities. The spatial features of landscapes influence their distribution and abundance.

Two famous ecologists (Robert MacArthur and E. O. Wilson) were among the first to develop a theory that explicitly considered the role of space—the theory of island biogeography. It predicts that the number of species on an island (species richness) is a function of its size and its distance from a continent. A larger island has a lower extinction rate (more habitats and less competition) than a smaller one. An island closer to the mainland will have a higher colonization rate because immigration will be easier from the continent (where all species live). Thus, the closer and larger an island is, the greater its species richness.

A more recent viewpoint, metapopulation theory, builds on this original theory. It argues that in a heterogeneous region, the habitat suitable for a species will be fragmented into patches, so the regional population (the metapopulation) will comprise a group of local populations, each with its own dynamics (see chapter 20), but each subpopulation will interact with the others through occasional dispersal. Some of these local populations (source populations) will live in excellent conditions with lots of resources and will become large enough to force some individuals to leave and colonize other habitats. At the other extreme are sink populations, where individuals live but there are not enough resources, so that the subpopulation is constantly declining. Sink populations are maintained only if they are close enough to source populations to replenish their numbers.

Landscape ecology is a young science built on these two bodies of theory. In this field ecologists define three elements of space: the matrix, which is the dominant background element; the islands of livable habitat distributed in the matrix; and corridors, habitats that link the islands.

Each species will perceive its habitat differently: the same matrix will not have the same effect on different species. In the city, the matrix for our butterfly is the urban structures. The islands are the parks, the forest patches, and the industrial wastelands. These islands are connected by corridors: trees along the streets, vegetation bordering railroad tracks, the banks of rivers.

Landscape ecology seeks to explore the complexity of landscape patterns. Ecologists interested in the urban environment have their work cut out for them: the size, isolation, and connectivity of urban islands, as well as the quality of urban habitats—often polluted, disturbed, or invaded by exotics—create a complex web of interactions that can be hard to interpret when all are enmeshed in a heterogeneous matrix.

20

ENOUGH ALREADY!

A legend is told in India about the Brahman Sissa, who invented the game of chess to entertain his sovereign, King Belkib. The overjoyed king asked what Sissa would like as a reward, to which the philosopher astutely replied: "I would like you to place a grain of wheat on the first square of the chessboard, two grains on the second, four on the third, sixteen on the fourth, and continue doubling the amount until all sixty-four squares are covered." Belkib agreed to do this, though he laughed at the apparent modesty of this request. Obviously he did not understand the principle of exponential growth as well as the clever Sissa.

When it came time to pay up, King Belkib realized to his horror that there wasn't enough grain in his entire kingdom: he would have needed 18,446,744,073,709,551,615 grains—the number harvested from all the wheat plants on the whole planet for about five thousand years—evidence of the impressive effect of exponential growth! (The king's solution was to cut off the brahman's head.)

Most animal populations have the potential to grow exponentially, mainly because reproduction is multiplicative; if one woman gives birth to two girls, and each daughter does the same, that makes four female grandchildren, and sixteen female great-grandchildren were the trend to continue, and so on.

However, although most populations without predators do undergo exponential growth, this growth will inevitably be followed by a crash. When the population surpasses the **carrying capacity** of its environment before realizing it is out of food, resources start to become limiting and it can no longer grow. At this point death rates increase and birthrates become lower as famine sets in. The population will then decline until it reaches a number that allows resources to be replenished. How and why these processes occur in natural populations is the basis of the study of **population dynamics**, founded on the principles of **demography**.

SURROUNDED BY GEESE?

The Canada goose (*Branta canadensis*) that ranges across the entire continental United States and Canada is a good example of a population that is growing exponentially. Today it is hard to believe that not long ago it was an endangered species. During World War II, hunters in central North America despaired of ever hunting it again: excessive hunting at the turn of the twentieth century, as well as the destruction of wetland habitat, almost caused its extinction.

A few small subpopulations were found in the early 1950s, and several reintroductions were attempted in the central and eastern United States, leading to a substantial increase in the overall population. During the first Canada goose reintroductions, wildlife managers wanted to introduce a new subspecies, the giant Canada goose (*Branta canadensis maxima*), in an effort to maintain wildlife diversity. They were also enthusiastic about a new game species for hunters. These geese were bigger, were present year round, and could be found close to the cities where most sport hunters lived.

Today giant Canada geese couldn't be doing much better. Their numbers increase every year, worrying many wildlife managers. Subpopulations are now year-round residents in southern Canada

and the northern United States. Some remain migratory, and you can see and hear flocks overhead in the spring and fall. Migratory geese nest in northern Canada, where the harsh environment prevents exponential population growth.

In contrast to their migratory brethren, the resident geese are very tolerant of humans, so they live with us year-round. Furthermore, these opportunist herbivores take advantage of many human-induced changes to the environment. They find golf courses and parks ideal habitat: lots of grass to graze on, ponds to bathe in and retreat to when disturbed, and a near absence of predators (including hunters).

In addition to the well-kept and fertilized grass in cities, these geese often find abundant food in the fields of corn and other grains that may be just a few wing beats away. And as if supplements were necessary, geese are often fed by humans who appreciate having "wild animals" near where they live.

A FRIGHTENING SUCCESS

No one predicted how well this subspecies would do in the urban environment. Trying to remedy their growth, many managers raised hunting quotas in the 1980s to increase the death rate. Unfortunately this had little effect on populations because the individuals most vulnerable to being killed were juveniles, and demography shows that increasing the death rate of the nonreproductive members of a population doesn't affect the overall birthrate when there are plenty of adults.

The giant Canada goose causes all sorts of problems. The biggest one for most of us is that they leave copious feces—perhaps you've experienced firsthand the mess left after hundreds have gathered in a park for a week. Something similar happens when a large number occupy a lake or pond: the water quality can be affected to the point where diseases can spread and swimming must be banned. There is also a great danger of collisions between geese and airplanes. In 2009 a flock of Canada geese was sucked into the jet engines of US Airways Flight 1549 taking off from LaGuardia Airport in New York, forcing it to land in the Hudson River (successfully). Finally, they damage the crops they feed on.

In many places, data show that the giant Canada goose population is growing exponentially. The problem is that they likely are still far from reaching their carrying capacity. Although we have found some ways to deal with them, we will probably just have to learn to live with these noisy, messy neighbors.

VULTURES GALORE

Farther south, another troublesome bird has reared its mottled head. The black vulture (*Coragyps atratus*) is a native of the southern United States, and its range extends to South America. Recently its population has been growing alarmingly in some regions such as Tennessee. Like other animals that do well around humans and cities, these vultures commonly feed at garbage dumps (see also chapter 24), easily fueling their population growth.

These vultures (unlike turkey vultures) are gregarious and roost in large groups, causing problems for power companies as well as city dwellers because the accumulation of their feces short-circuits power lines. Around cities their droppings cause even more problems, piling up on trees, lawns, and buildings and contaminating water supplies. They are attracted to human products and will tear at or eat roofing material, car upholstery, windshield wipers, latex window caulking, and plastic flowers.

Control attempts use light, sound, and water to harass them and scare them away, but as animals often do, they habituate to these tactics and a new approach is needed. Stretching wire or monofilament over structures may keep them from perching, but this is obviously limited on a large scale. Population management may take more drastic measures including culling, a controversial approach now being used in some areas of the United States.

GO FORTH AND MULTIPLY

It's not just goose and vulture populations that grow exponentially. Since our own species, *Homo sapiens*, first appeared more than two hundred thousand years ago, we have spread across the planet. But it is mostly since the Industrial Revolution that the human population has grown exponentially. It took several millennia to reach one billion individuals but only another two hundred years to reach six billion.

The explosion of the human population has also greatly affected the demographic growth of other species. For example, white-tailed deer, gulls, coyotes, and raccoons, to name just a few, have benefited from our expansion. Although we often hear that the destruction of wildlife habitat has led to a species' decline, we sometimes forget that other species feel a little too comfortable in their human-created habitats.

Madeleine Doiron and Mathieu Beaumont

CAFE CONVERSATION

Discuss a troublesome species in your neighborhood or region. What do you think has enabled this species to be so dominant? Consider the behavior and life history of the organisms as well as the opportunities the urban environment might have provided.

DEMO, DEMO, DEMOGRAPHY!

Demography is the study of populations and the processes that lead to their growth and decline. The basic factors affecting population growth are birth, death, immigration, and emigration. Immigration and birth are the only ways individuals enter a population, while emigration and death are the only ways they can leave it. The difference between these processes changes population size.

A demographic explosion can occur whenever access to resources no longer limits a population. In this case the addition of individuals (by birth and immigration) is always greater than the loss (by death or emigration), and the population will grow continuously for a time. But as the density of individuals grows, competition between them will also increase, so eventually food resources become scarce. Then death rates rise and birthrates decline, making the population smaller.

21

DARWIN'S SWEET TOOTH?

A hungry but indecisive freshman contemplates the candy display at the corner store near his new university. Next to the major brands of chocolate (the ones television ads show being relished by fit young people) the student notices some strangely named chocolate bars: White Skunk, Fat Monk, Sweet Olive . . . Why, he wonders, can you find such a variety of chocolate in this big-city shop when his small hometown has only a few major brands to choose from? How can these unknown brands survive in a world dominated by the well-oiled marketing machinery of the large candy companies?

Such questions are very like those ecologists ask when they wonder why the biological world is so diverse and why its diversity varies from place to place. Since the start of their science, ecologists have pondered the great diversity found in plants, insects and other animals, and microorganisms. They also ask, for example, why a section of the Alaskan boreal forest the size of a football field may contain only a dozen species of trees while the same surface area in a tropical rain forest would rarely have fewer than two hundred.

These questions are essential to ecology, which strives to explain the abundance and distribution of species in different environments. Given the similarity of the questions, maybe the same concepts ecologists use can explain the distribution of rare and exclusive chocolate brands in cities and in small towns.

There is currently an important debate among ecologists about biological diversity. Two diametrically opposed theories are in play: niche theory and the neutral theory of biodiversity. Let's examine what's behind this quarrel.

NICHE THEORY

Since the dawn of ecology, researchers have believed that competition between individuals for resources is the fundamental explanation for both the distribution of species and their differentiation. From this assumption emerges the hypothesis of **competitive exclusion**, which predicts that when two similar species compete within the same community, the weaker competitor will eventually be eliminated by the stronger.

This hypothesis is fairly intuitive and can equally well be applied to our chocolate species (brands) in a marketplace where the number of consumers is limited. Those rare and exclusive brands with superior taste (for connoisseurs) or lower price (for poor students) will have a competitive edge over other brands and are predicted to outcompete them and dominate the market. This is what we would expect if niche theory applies. According to this theory, living species (or brands of chocolate) have evolved to be maximally efficient in a given environment so as to outcompete other species that require exactly the same conditions.

The greater the diversity of conditions in a given environment, however, the more difficult it will be for a single species to be the most efficient. This is captured in the principle of **biogeography**, summarized by the great ecologist Robert MacArthur as "jack of all trades, master of none." His theory predicts that an increase in the diversity of environmental conditions and resources will make more niches available, translating into more species.

If a brand of chocolate is a species, the subset of consumers it most appeals to defines its niche. The city includes people with a wide va-

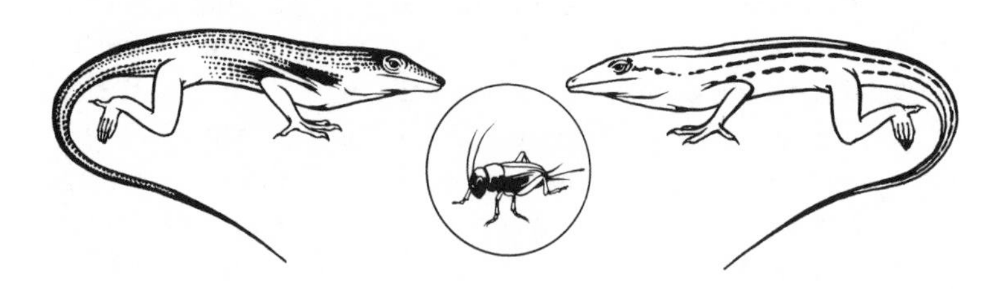

riety of tastes and incomes, creating a large diversity of niches. This allows the smaller and more exclusive brands to survive despite the aggressive advertising and marketing of the large candy companies. But in a smaller town, where there are fewer consumers, whose tastes and incomes are probably more homogeneous, it is almost impossible for an olive-flavored green chocolate to find enough consumers to survive.

THE NEUTRAL THEORY OF BIODIVERSITY

Recently a different viewpoint has called niche theory into question. Neutral theory was first proposed in the early 2000s by two separate researchers: Stephen Hubbell, based on his studies of tropical forests in Panama, and Graham Bell, an evolutionary biologist at McGill University, based on microorganisms and theories from the study of evolution. Not only did this new theory question the concept of the niche, it even questioned Darwin's theory of natural selection.

Imagine that most chocolate lovers are "neutral" in their preference. Standing in front of the chocolate display, your choice of candy bar would be random, dictated by neither price nor taste (which might actually be true for many customers).

The only factor that would affect your choice in this case is the relative abundance of each brand. The more common a brand, the more likely you are to choose it at random. In the long term, this way of choosing chocolate would have repercussions on other brands because rarer kinds would eventually be forgotten and excluded from the market. The storekeeper's decision on which to stock would depend on the quantity received and the distribution strategies of each

candy company and would have nothing to do with competition to satisfy a discerning consumer.

According to neutral theory, all species living on Earth would be organized in the same way. The composition of the flora and fauna would be determined randomly by the relative abilities of species to disperse quickly and in large numbers, with environmental tolerance playing no role. In the long term, rare species that do not disperse well would always be eliminated. The only way a community could compensate for the loss of rare species is through immigration from neighboring communities. The more new species (or brands of chocolate) immigrate into a local community, the greater will be its diversity.

In such a scenario, rare species would be immigrants subject to successive local extinctions and reintroductions. Common species would be more stable. This explains why the large candy companies can keep producing the same brands for a long time (they overwhelm us with advertising and inundate vendors' shelves with their chocolate), while small specialty chocolate makers tend to have a higher turnover (some disappearing while new ones show up all the time). According to this theory, if more candy stores opened in either a big city or a small town, that area would also see an increase in the diversity of chocolate because there would be more room for new brands to "immigrate." Simply put, big cities have more types of chocolate because they have more chocolate factories and stores.

THE CHOCOLATE MELTING POT

Paradoxically, the two theories explain the same patterns of diversity and make the same predictions but rely on opposite principles. Many researchers believe that both theories are true and which one applies will depend on the conditions of the system under study. Over time, a unified theory is likely to emerge for the science of ecology.

This debate is more than a simple spat between a few ecologists. An improved understanding of how the theories apply to species distribution would allow environmental managers to better evaluate the consequences of new species introductions, conserve rare and endangered species, and assess the effect of ecosystem fragmentation linked to habitat loss.

In either case, the next time you buy a chocolate bar, ask yourself, Am I "niche" or "neutral"?

Dominique Gravel and Christian Messier

CAFE CONVERSATION

Discuss the two theories this chapter presents in the context of human behavior and merchandise distribution. Which strategy best explains the diversity of products we find in different parts of our city or between cities? Could this help your business or a friend's?

THE THREE SCALES OF BIOLOGICAL DIVERSITY

Biological diversity varies greatly between regions on our planet. Globally speaking, diversity is so large that its extent is still largely unrealized. Some scientists estimate that from ten million to one hundred million species remain to be discovered! New ones are found almost every day. But remember that planetary diversity changes constantly both through the loss of species not well adapted to their environment and from speciation events that produce new species.

We also know there have been several large waves of species extinctions caused by natural disasters, including the impact of large meteorites or the simultaneous and massive eruption of many volcanoes. Since the origin of life on Earth, several million species have been produced.

Ecologists describe species diversity at three scales. At the planetary scale we know that diversity increases from the poles to the equator, partly because environmental conditions in the tropics are more suitable for life, more stable, and less affected by periodic glaciation. At the regional scale, Earth's geology is dynamic and tends to geographically isolate species by forming mountains, isthmuses, and so on, contributing to the formation of new species. Finally, at the scale of a community, biological diversity is linked to the diversity of environmental conditions and the attributes of species (niche theory) or to random events and the dispersal ability of different species (neutral theory).

22

A BREATH OF FRESH AIR

A pleasure to behold, peaceful oases of green in the gray city, the lawns, trees, and shrubs that soften the urban landscape also perform several services for this ecosystem that is so overrun by people. We often overlook one of the most important jobs all this greenery does, one that involves passing millions of cubic meters of air in and out of the stomates of urban plants each year. Let's explore.

City air is not always the most breathable. Over the past few decades, concentrations of many air pollutants and particulate matter (very fine dust) have been on the rise globally. This pollution is often visible as a yellow-brown fog above cities, known as **smog** (a contraction of "smoke" and "fog").

Among the offending gases, nitrous oxides and sulfur oxide, commonly released by vehicles and large industries, are the most important. On hot, sunny days these gases can produce ozone, but even plants, in particular coniferous species, can pollute the air with volatile organic compounds (VOCs) that help make up smog. And

cities produce another major group of pollutants, including macroparticles of dust formed in part by the natural erosion of concrete. Such pollutants have variable effects on our urban environment, but they are always harmful.

Many scientists think that the rise in very fine dust particles as well as coarser atmospheric dust is one of the major causes of the increase in allergies among the young and of breathing difficulties among older city dwellers. Many microbes that our immune systems sometimes overreact to are associated with these particulates.

The many harmful effects of atmospheric CO_2 are more indirect but are now well understood. The main one is as a greenhouse gas, responsible for global warming, an issue we'll return to shortly.

The health risks associated with nitrous and sulfur oxides as well as ozone are more direct. The chemical structure of these compounds lets them produce very aggressive molecules called **free radicals**. In living organisms, these molecules react with living tissue to attain a more stable state. Unfortunately the result is **oxidation**, which irreparably damages tissues of all city-dwelling organisms and can cause diseases like cancer. Despite all this, it could be worse.

TAKE A DEEP GREEN BREATH

Vegetation growing in the city helps remediate declining air quality. Grass, shrubs, and trees work quietly away in our cities' parks and gardens, bettering the air quality just by living where they do.

Plants greatly reduce the particulate matter in the atmosphere. Their leaves are covered with a layer of protective wax called a cuticle, produced by the epidermal cells. Impurities cling to this cuticle as the wind blows them there, then rain cleans the leaves, ultimately carrying the dust down into the sewers. Nothing could be simpler for plants than reducing carbon dioxide, because they use it the same way we use oxygen. Air enters small openings in the leaves called stomates. Plants require CO_2 for photosynthesis, a chemical process that uses sunlight as energy to produce the sugars they need to grow (see chapters 1, 2, and 10). A by-product of photosynthesis is oxygen, essential to humans and other animals, which the stomates release back into the atmosphere. Everyone benefits! Although the quantity of atmospheric CO_2 fixed by plants on Earth annually is enormous, much of it is later released by decomposition (see chapter 1) when leaves fall or the plant dies. Thus we can't rely solely on plants to reduce the level of atmospheric carbon to acceptable levels and avoid climate warming altogether.

The final way plants help clean our air is by being incredibly effective at trapping free radicals. As they capture CO_2 through their stomates, they also absorb large quantities of free radicals from the air. Plant tissues contain many antioxidants that quickly neutralize them before they can attack the plant's cellular tissues, transforming them into innocuous, nonreactive molecules.

Urban vegetation, particularly nonconiferous species, contributes in this way to reducing the harmful free radicals in the air. But there's no free lunch: this battle is costly for the plants, which become more fragile and sensitive to physical damage as well as to insect and microbe attacks. Thus it's crucial to protect our urban green spaces so plants can carry out their precious air filtration and carbon reduction.

GREEN AIR CONDITIONING

Urban vegetation also benefits the microclimate of cities, which are up to 41°F (5°C) warmer than the neighboring countryside. Higher city temperatures are mainly caused by a concentration of the sun's rays reflecting off building surfaces. Furthermore, the omnipresent asphalt and concrete store heat effectively and continue to release it for a long time. Thus the presence of vegetation instead of concrete directly reduces the temperature in cities.

To understand why, let's consider the city's energy balance in terms of its **thermal energy** content. There are energy sources that emit heat and energy sinks that capture it. In the city, sunlight is the major energy source, and all objects including living organisms are energy sinks. But not all objects and organisms absorb heat equally. An object's ability to absorb solar heat is its **albedo** (Latin for "white"). The closer an object's albedo is to zero (the value of a completely black object), the more solar energy it absorbs; the closer it is to one (a completely white object), the more sunlight it reflects.

Concrete, asphalt, and metal do not reflect solar energy well; they absorb most of it. These materials have low albedo and therefore heat up and become secondary sources of heat. It's easy to understand

why urban temperatures are so much higher: there's an overabundance of both primary (e.g., sunlight, car engines) and secondary heat sources.

In rural environments, the dominance of vegetation dissipates solar heat, lowering the environmental temperature. Trees have a moderate albedo index; they reflect little energy, but they do not warm up either because they can dissipate the energy. The mechanism is fairly simple: they cool off when they transpire, much as we do when we sweat.

Plants provide incomparable service to city dwellers. Consider thanking the next one you pass. You'll need to get close, not so much because plants are hard of hearing (they're completely deaf) but because they'll appreciate the CO_2 you release when you talk to them. A delicious reward.

Sylvain Delagrange and Frank Berninger

In addition to looking at lichen abundance (chapter 18) to assess how desirable a neighborhood might be, look at a satellite view of your city on the Web (as in the activity for chapter 19) and determine the amount of greenery in that area. A good compromise might be between nice neighborhood cafés and the quality of the air you'd breathe.

23

SUCCESSFUL LITTLE BIRDS

As we saw in earlier chapters, some species introductions lead to dramatic population explosions. The urban story of the house sparrow (*Passer domesticus*) merits its own chapter. It's true they're cute bouncing between the pigeons as they try to capture a few crumbs. It's certainly true that their joyous calls in the morning make the city more cheerful. But as the ecologist in you has guessed, the house sparrow has a sinister side.

The house sparrow has been introduced to North America several times, and each time people have recorded the name of the person responsible, the date, the location, and the number of birds released. We know, for example, that in 1852 Nicholas Pike imported fifty sparrows from England and released them in Brooklyn. That was the first successful introduction into North America, but until 1880 there continued to be many more reintroductions, mostly in the New England states.

Enthusiasm for the house sparrow stemmed mainly from colonists' nostalgia for the animals of home and from the hope that sparrows would help control insects. Sparrows were purposely introduced around the world, and these little birds are now abundant on all continents except Antarctica.

ORIGINS OF THE URBAN SPARROW

What's the origin of this animal that has become so cosmopolitan? Sparrows belong to the family Passeridae and are closely related to the weaver finches of Africa, Eurasia, and Oceania. Their original distribution was large, including Eurasia and northern Africa but not the Americas. The male house sparrow can be recognized by his prominent black bib, a black patch of variable size on his throat.

Why were sparrows so successfully exported? Certainly they benefited from the initial sympathy of their human benefactors, who made sure they were well fed, protected, and supplied with nest boxes. But as we have already suggested, those early settlers should have been warier. House sparrows are incredibly fertile, producing several clutches each summer. They also possess an amazing morphological plasticity. For example, the size of individuals increases from south to north, and the extra accumulation of fat and water helps the birds resist cold winters. They also have an incredible power to learn and innovate: that is, a large ethological capacity (from Greek *ethos*, "manners"). In New Zealand sparrows have learned to operate the automatic sliding doors of a bus station so they can forage for food inside. In many other cities, sparrows take up residence in shopping malls, living there in comfort all year round.

BIG PROBLEMS WITH LITTLE BIRDS

A major reason this exotic species has an impact on the native fauna of North America is that the birds are cavity nesters. Because they don't migrate, they begin their reproductive season earlier than many indigenous birds and take over the limited number of nest cav-

ities that would otherwise serve native species. Furthermore, sparrows are very territorial and ferociously defend their nests. They can also be bullies, evicting species like the eastern bluebird (*Sialis sialis*) and the purple martin (*Progne subis*) from their own nests.

The house sparrow has also had an effect on humans' houses, economy, and health. By 1899, after the many successful introductions, people started to call them "winged rats" because they build nests in vents, fissures, gutters, and crevices of walls, causing damage and blockages. They can injure young vegetables in gardens and will gorge themselves on newly planted seeds. They also do considerable damage to fruit trees and cereal crops.

Several methods have been used to control sparrow populations, from poisoning to hunting. Some methods are more creative than others. Toward the end of the nineteenth century, the classified ads in some newspapers incited people to capture live sparrows for target practice. Not the most humane treatment, and unlikely to be tolerated these days. Then again, none of the methods used have had more than a slight effect on the phenomenal abundance of sparrows in North America. Once again, an exotic species seems to have become solidly entrenched in its new environment.

Phoenix Bouchard-Kerr and Luc-Alain Giraldeau

CAFE CONVERSATION

Starlings, pigeons, and house sparrows were all introduced to North America by Europeans. Dozens of other species of birds were also brought over, but they are nowhere to be seen. Every year birders find stray European birds on the eastern coast of North America, blown here by some major storm. Sooner or later some of these may become as successful as starlings or house sparrows. Do you think some of the species currently living in North America could have been part of such "natural invasions" thousands of years ago? How could we find out?

FIVE OTHER IMPORTANT INVADERS

African Honeybee (*Apis mellifera scutellata*)

Origin: Africa.

Invaded regions: Introduced to Brazil and spread across South and Central America into the southern United States.

Vectors and pathways: Introduced to augment the honey industry in Brazil. The rest of the expansion took place by the bees' own movement.

Impact: When it crosses with the domestic honeybee, their offspring become far more aggressive (though there is no increase in the toxicity of the venom). There have been many deadly attacks on domestic animals and humans. They also produce less honey.

European Starling (*Sturnus vulgaris*)

Origin: Europe, southwestern Asia, and northern Africa.

Invaded regions: North America, South Africa, New Zealand, and Australia.

Vectors and pathways: Introduced to the United States by nostalgic colonialists and brought to New Zealand to control insects.

Impact: Compete with native fauna for food resources, evict other cavity-nesting birds, damage crops, spread diseases (zoonoses).

Asian Long-Horned Beetle (*Anoplophora glabripennis*)

Origin: China and Korea.

Invaded regions: North America, Great Britain, and Austria.

Vectors and pathways: Accidentally introduced on wooden pallets used for maritime shipping.

Impact: Attack and kill many species of deciduous trees, most notably maples.

Black Rat (*Rattus rattus*)

Origin: India.

Invaded regions: The entire world.

Vectors and pathways: Accidentally introduced as stowaways on transoceanic ships.

Impact: Eat pretty much anything edible (and more), have caused the extinction of many species of birds, small mammals, reptiles, invertebrates, and plants, especially in fragile island ecosystems.

Zebra Mussel (*Dreissena polymorpha*)

Origin: Southeast Russia (Ponto-Caspian region).

Invaded regions: North America, Great Britain, and Europe.

Vectors and pathways: Accidentally introduced in ballast water from ships.

Impact: Damage boats and harbors, block pipes in water treatment and power plants, grow in monocultures, excluding native bivalves.

24

OPPORTUNISTIC GULLS

The ring-billed gull (*Larus delawarensis*) exemplifies a recurring wildlife management problem caused by three typical factors: the biology of the species, its habitat, and its positive response to humans. We don't know how many gulls there were in North America when the European settlers first arrived in the seventeenth century, but we do know that toward the end of the nineteenth century and into the twentieth the species was heavily exploited for its eggs and for its feathers, which decorated many a lady's hat, and the gulls became extremely rare. As a result, in 1918 ring-billed gulls were added to the Migratory Bird Treaty Act between the United States and Canada. Since then the population has gradually reestablished itself. By about 1930 gulls were again regularly seen during the migration period. Today it is estimated that the North American gull population is about 1.7 million strong. Government agencies do regular censuses, providing essential information about its population dynamics.

DYNAMIC DYNAMICS

Ring-billed gulls begin to reproduce at three years old and live to be ten to fifteen. They nest in colonies ranging from a few hundred individuals to a few million. Each female lays three eggs in a nest on the ground, shaped like a cup and made of twigs and other materials. Gulls are fond of living on islands, which helps protect them from mammalian predators.

Gulls are incredible opportunists, giving them many advantages and allowing them to adjust readily to new conditions. In addition to seeking out islands, some now nest on flat roofs of buildings. Although this new behavior is still not well established, where it does occur people try hard to discourage the habit before it spreads within the population leading to population explosions. Given the wide availability of such habitats in some cities, the chances of its spreading are frighteningly high.

Gulls are model parents and regularly relieve each other in incubating their eggs, which they do for just under a month. The newborns are semiprecocious: they leave the nest several hours after hatching, but their parents feed them until they fledge at about thirty-five to forty days old. The long life span of the ring-billed gull and its high nesting success partly explains the 10 percent annual population increase observed at the end of the twentieth century. But another factor has also contributed to making this gull so successful.

LET'S DINE IN TOWN

Gulls often find food close to their colonies, but they can easily fly up to 37 miles (60 km), giving them variety of other opportunities. Household garbage constitutes about 40 percent of the diet of gulls nesting close to larger cities. Many regularly visit dumps, despite laws that operators bury garbage quickly. Less patient birds don't bother waiting for the garbage to arrive at the dump. They patrol residential communities on garbage pickup day and rip open bags with their sharp beaks. In many city centers, they frequent dumpsters and garbage cans with poorly fitting lids, especially near restaurants.

As if they needed it, some people even supplement gulls' diets, usually because they enjoy interacting with them. Many birds wait near picnic areas; they rapidly find places where there are handouts and regularly beg for food, sometimes aggressively.

Add to this the natural diet of ring-billed gulls, which is also variable: earthworms, insect larvae and flying insects, fish, and small mammals. Because its feeding habits depend largely on what it encounters, ecologists refer to this species as a **generalist** and **an opportunist**. We now know that populations exceed the natural carrying capacity of their natural environment largely because they are so well supplemented by the garbage of our cities.

A SMALL SURVEY OF NUISANCES

Gulls carry various pathogens including *Escherischia coli*, potential sources of disease that they spread through their droppings. Other bacteria they carry include *Staphylococcus*, *Salmonella*, and *Listeria*. It's best to avoid petting gulls; luckily they aren't inclined to demonstrations of affection.

Provided one follows government regulations for water quality, the quantity of bacteria from gulls that you could ingest while swimming is unlikely to harm you, yet this doesn't mean gulls don't contribute to the degradation of beaches. But if beaches are monitored and you respect "closed" signs, the chance that gulls will infect humans is minor.

The droppings that accumulate on statues, buildings, car roofs, and waterfront property cause damage and are expensive to clean up. As with Canada geese, gulls' presence near airports is also a major problem (see chapter 20). According to Transport Canada, the greatest number of declared collisions with planes has been with ring-billed gulls.

MORE FEATHER-TOPPED HATS, PLEASE!

Pyrotechnic and acoustic techniques, including exploding cartridges, detonators, and recordings of distress or alarm calls, have had limited success in dissuading gulls from using sites where they aren't wanted. Ring-billed gulls habituate very quickly to such strategies. Keeping falcons at a site is effective at scaring them away, but that approach is labor-intensive and expensive. Physically excluding birds by stretching nets over a section of a beach or other public place can work, but for practical reasons this can't be done everywhere.

In any case, these methods simply move the problem elsewhere. More radical control methods include reducing birthrates by destroying eggs or spraying them with mineral oil to prevent hatching, but such tactics must be implemented many times in a season to affect the overall population, since the birds lay many clutches each year. Killing birds directly requires a permit and is expensive and often controversial.

The effective management of wildlife populations always relies on knowing the basics of applied ecology. When data are incomplete or fragmented, management methods must be limited and put in place one step at a time to avoid irreversible errors. But with a good understanding of a species and its interactions with the environment, more drastic interventions are safe. For the ring-billed gull, the ultimate solution is likely a more global approach that includes reducing garbage at its source, managing waste better before and after collection, applying a variety of dissuasion and exclusion methods, and not least, discouraging people from sharing their picnics with gulls.

Jean-François Giroux

CAFE CONVERSATION

Should humans intervene as "predators" when a natural population gets too abundant for our taste, or should we put our efforts into increasing its natural predators? What advantages or problems do you see with each approach?

25

URBAN ROOTS

In previous chapters we have learned about trees' ingenious water management (chapter 2), the history their rings record (chapter 17), and the quiet, unselfish ways they make the city more livable (chapter 22). But how do trees survive in the city?

Often isolated from neighboring trees by an inhospitable matrix of concrete and asphalt, life is not easy for an urban tree. Its soil is trampled on, its bark vandalized or scarred by passing snowplows. It has little space to grow. People trim branches to keep them from touching power lines, traffic lights, and street signs. A tree also

has to deal with atmospheric pollution and often with inadequate sunlight.

Underground the story is even more tragic. In the forest, a tree's roots can spread over an area several times larger than its crown to find the nutrients for growth. In the city, where soil is hard to come by, trees planted on the borders of sidewalks are stuck between the sewers and the foundations of buildings and, more often than not, have less than four cubic yards of soil available—very little considering their crown volume can easily reach a hundred cubic yards. Moreover, what little soil there is may be poisoned by all sorts of toxic substances. Soil contamination is at its worst in northern climates when the snow melts during the spring thaw and the salt spread all winter to de-ice streets and sidewalks becomes soluble, reaching up to six hundred times the acceptable levels.

City soil also lacks aeration because it is covered with asphalt and is compacted and nonporous. Oxygen is crucial for the survival of insects, microorganisms, and bacteria that aerate soils, as well as for root metabolism and for mycorrhizae, those happily serving symbiotic fungi (see chapter 2). All this subterranean life works to convert mineral and organic substances in the soil into nutrients that trees assimilate and use. But when oxygen levels are too low, these organisms cannot survive, and **anaerobic bacteria**, those that work in the absence of oxygen, have the upper hand. Instead of producing useful nutrients, these bacteria produce acids and methane, toxic to tree roots.

Together these unfavorable conditions affect the health of trees so much that their life span is reduced to about five years along downtown sidewalks—a very short time when we consider that in the forest many species normally live two centuries or more.

The sad story of the urban tree is made sadder by the attitude of many city dwellers. How many people plant a tree without considering its size at maturity? How many complain when fruits stain their cars or when they have to rake leaves from their swimming pools in the summer and their lawns in the autumn? The unappreciated tree would have been a worried sapling had it known it was so likely to be cut down. We expect trees to be pretty but not to get in our way—the fate of the "tree as object." This view extends even to many city bylaws, which often treat trees the same as bus shelters and traffic lights: as "movable objects."

DON'T MISS THE FOREST FOR THE TREES

Trees that grow in the city are part of a green mosaic called the urban forest, providing life to this seminatural ecosystem dominated by humans. The urban forest harbors many other living organisms including herbaceous plants, vertebrate animals, insects, and microoganisms. The soil, air, sun, and water form the **abiotic** (nonliving) structure of this ecosystem.

The tree is the central point around which other organisms interact. Trees captive inside a small rectangle of ground within a sidewalk will have weak interactions with other living things, though interaction can be considerably greater in parks or in more naturalized parts of the city where trees are less confined.

Considering the urban forest as a living ecosystem allows for better management of green space. Many cities have recently launched naturalization programs that work toward this goal. These cities have put away the lawnmowers and grass seed in favor of a cover of natural vegetation in less frequented areas of parks and along bicycle paths.

Diversity is a key part of the complexity of any ecosystem, because it ensures more interactions among species (see chapter 4). Low levels of diversity, in conjunction with urban stress, make the urban forest susceptible to insects and diseases. For a voracious or

disease-carrying insect, nothing is easier, after all, than to hop from tree to tree of the same species along a city avenue.

Dutch elm disease is an excellent case in point. Introduced to North America about 1920, this disease is deadly for native elm trees. It is caused by a fungus (*Ophiostoma ulmi*) spread by small bark beetles as they move from tree to tree. In many cities the result has been disastrous: many streets lined with a monoculture of majestic elms saw all their trees destroyed within a few decades.

Viewing the urban forest as an ecosystem helps us better understand its ecological functions. Physiologically, it ameliorates air quality, regulates temperatures, reduces winds, controls erosion and improves drainage, and favors the conservation of high-quality soils. Higher in the food chain, the urban tree plays a role similar to that of its forest brethren; many species get both shelter and food from city trees.

Certain important ecosystem functions, however, are absent or greatly reduced in the urban forest. Take the nutrient cycle (see chapter 1). In a forest ecosystem, the nitrogen tree roots absorb comes from leaves that microorganisms decompose into humus. After being used by the plant, nitrogen returns to the soil the next autumn when the leaves fall. In the city the cycle is incomplete because the leaves are collected and removed and because the base of the tree may be covered by grass or, worse, by concrete or asphalt.

Without a doubt, life is hard for urban trees. But by choosing the most resistant individuals, planting them in the best places, giving them enough space and resources, and maintaining a diversity of species, we improve conditions for the urban forest and ensure that it survives to benefit us all.

Isabelle Aubin and Benoît Hamel

Type "ecosystem services" into a Web search engine and look at the list that comes up. Discuss how many of these services urban trees provide. Why do you think urban planners don't do more to encourage planting trees? What could you do about this?

WHAT IS AN ECOSYSTEM?

The term ecosystem was first used in 1935 by the British ecologist A. G. Tansley. The concept quickly became a central theme in the study of ecology.

An ecosystem is a biological system made up of its associated species, including all their interdependencies, in an environment that allows for maintaining and developing life. In effect, an ecosystem is to the organism what an organ is to the cell. There are five major attributes of ecosystems.

1. Structure. To allow a terrestrial ecosystem to function, there must be a community of living organisms, soil, air, an energy source (sunlight), and water.
2. Energy exchange. The ecosystem allows for exchange of matter and energy between the physical environment and the living community. From this perspective, a living organism is primarily an accumulation of solar energy living in association with the chemical elements stored in the soil and the atmosphere.
3. Interactions and interdependencies. The components of an ecosystem are interconnected, somewhat like a spiderweb. These connections are sometimes so direct that a change in one component will affect all the others.
4. Complexity. Because its biological components are highly integrated, an ecosystem is a complex system.
5. Changes. An ecosystem is never static. It is in constant movement, with constant flux between components. This lack of a status quo allows for adaptation.

CONCLUSION

Ecology is the science that deals with the interactions between organisms and their environments. Although it is not the same as environmentalism, a good grounding in ecology is required for many of the problems environmentalists tackles, especially those dealing with conservation and management, or even restoration. Ecology provides knowledge of the basic interactions that environmentalists use to set baseline conditions. While an ecologist might ask how or why two species interact (e.g., koalas and eucalyptus trees), an environmentalist might want to know about such a relationship to inform decisions that might affect either species. For example, cutting the eucalyptus forests may affect koalas, or culling koalas may affect eucalyptus forests or other herbivores. If you've tried some of the café conversations and activities suggested at the ends of the chapters, you've seen that conservation and environmentalism can be linked with basic ecological principles.

As we have seen, there are several subfields of study within the science of ecology. Sometimes the focus is on interactions between individuals of the same or different species, and how these habits or behavior came about by evolution is the domain of **behavioral ecology**. Considering the internal or physicochemical responses of particular organisms to their environments is the domain of **physiological ecology**. Some ecologists are interested not in individuals but in populations, and they study **population ecology** to understand, for instance, why a particular population might fluctuate more or less under certain environmental conditions or in the presence of another species such as a predator. When ecologists consider the interactions between many species and their environments, they are studying **community ecology**. Finally, when the focus is on how some key flows of nutrients or energy occur in a certain area, the field of study becomes **ecosystem ecology**. Within each of these subfields we can also consider **landscape ecology** as the study of such interactions at large spatial scales. In this book we have seen examples of each of these types of ecological research. Now that you have a more global view, we challenge you to return to each chapter and consider which branch of ecology it most likely fits. You can also ask yourself whether it makes sense to talk about, for example, an ecological car or house.

Thank you for joining our tour of ecology in the urban environment. We hope you now have a better idea of what ecologists study and how they approach their questions. We also hope you have learned a bit more about the ecological wonders waiting just outside your front or back door and sometimes even inside your house.

We encourage you to keep exploring the natural world, whether around your house, in your garden or the local park, downtown, or in a regional park. Use all your senses: listen for birds; smell the earth and the flowers; see the world around you from the tiniest ant to the biggest tree; touch the plants and animals (at least the nonpoisonous ones), and as you taste your food, think where it might have come from and notice its different flavors. All these activities will make you more aware of ecology and the ecological principles that are constantly in action. We humans also belong to this set of interactions. By understanding the ecology of the world around us we are poised to examine our place within it. We hope you'll continue to explore along with us.

Beatrix Beisner, Christian Messier, and Luc-Alain Giraldeau

GLOSSARY

Abiotic: Pertaining to nonliving chemical and physical factors in the environment that affect ecosystems. Abiotic influences include light, temperature, water, atmospheric or dissolved gases, temperature, topography, wind, fire, and soil conditions.

Adaptations: Functional traits of organisms that are evolved through natural selection and maintained in populations. These contribute to the fitness and survival of individuals.

Albedo: The reflecting power of a surface; more specifically, the ratio of the radiation reflected from a surface relative to the incident radiation on it.

Anaerobic bacteria: Bacteria that can survive with low or no oxygen.

Anemochory: The dispersal of plant seeds by the wind.

Angiosperms: The flowering plants. Characteristics include flowers, seeds containing a nutritious endosperm, and fruits that hold the seeds.

Anthropomorphism: Attributing human characteristics to animals or nonliving things.

Asci (sing., ascus): The sexual spore-bearing cells produced by the ascomycete fungi.

Autochory: Seed dispersal by physical expulsion from a plant (often explosively).

Autotrophs: Organisms that produce complex organic compounds (carbohydrates, proteins, and fats) from inorganic molecules by using energy most commonly from light (photosynthesis) and in some cases from inorganic chemical reactions (chemosynthesis). The most common autotrophs are algae and plants. They are also called primary producers.

Bacteria: Prokaryotic (no cell nucleus), usually single-celled organisms that are typically a few micrometers long and not visible with the naked eye.

Behavioral ecology: An evolutionary approach to the study of behavioral and life history decisions.

Biogeography: The study of the distribution of organisms and ecosystems in space and through geological time.

Bioindicators: Species used to monitor the health and integrity of an ecosystem or environment.

Biomass: The mass of living organisms in a given area or ecosystem at a given time.

Carrying capacity: The maximum size of a population that can be sustained by its environment.

Cellular respiration: The set of metabolic reactions and processes that take place within the cells of organisms and that convert energy from nutrients into adenosine triphosphate (ATP) and waste products.

Cognition: The central processing of information, such as learning, memory, perception, attention, problem solving, decision making, and language.

Commensalism: An interaction between organisms where one benefits while the other is neutrally affected (no harm nor benefit).

Community ecology: The branch of ecology that specializes in the study of species interactions at a variety of spatial and temporal scales.

Competitive exclusion: The extinction of one species by another through competition for resources.

Decomposition: The process by which organic material is broken down into simpler forms.

Demography: The study of dynamic populations and subpopulations, including their size, structure, and distribution as well as temporal changes.

Dendroarchaeology: The study of wood remains from old buildings, artifacts, furniture, musical instruments, and art objects using the techniques of dendochronology.

Dendrochronology: The method of dating by analyzing tree ring patterns.

Diapause: A delay in development of organisms in response to regularly recurring adverse environmental conditions.

Dicots (dicotyledons): The group of flowering plants whose seeds typically have two embryonic leaves (cotyledons).

Ecological niche: Most commonly considered to be the set of resources and conditions an organism or species requires for survival and reproduction.

Ecosystem ecology: The branch of ecology that studies the flows of energy and matter through both biotic components (organisms) and abiotic components (soil, atmosphere, water).

Entomopathogenic fungus: A fungus that initially colonizes an insect's body surface in the form of microscopic spores, then grows through the cuticle into its body cavity, usually killing it.

Estivation (aestivation): A state of animal dormancy in which the animal is inactive and has a lowered metabolic rate; used to survive conditions of extreme heat or aridity.

Evapotranspiration: The sum of evaporation and plant transpiration from a land surface to the atmosphere.

Exotic species: A species that has been intentionally or accidentally transported and released (usually by humans) into a region outside its original range.

Fitness: The ability to contribute surviving descendants to the next generation relative to the ability of other individuals in the population.

Flagellum (pl., flagella): A tail-like projection from the body of a cell that aids in locomotion and often acts as a sense organ (for temperature, chemicals).

Food chain: The linear succession of organisms (trophic levels) that eat prey and are in turn eaten by others starting with autotrophs (primary producers), which consume no one, and ending with those that are consumed by no other.

Free radicals: Molecules with unpaired electrons, often causing them to be highly chemically reactive.

Generalist: A species that tolerates a wide variety of conditions and can use many different resources (opposite of specialist).

Granivorous: Eating seeds.

Gregarious species: Animals that naturally occur in groups.

Guild: A group of species that exploit the same resources, often in similar ways.

Gymnosperms: The group of seed-bearing plants in which the seeds develop at the surface of the reproductive structures (naked seeds, not enclosed in ovules). Examples include conifers and cycads.

Heartwood: The wood in the center of trees, made up of dead tissue. Also called duramen.

Herbivores: Organisms that eat plant-based food; the primary consumers of autotrophs.

Heterotroph: An organism that cannot fix carbon (in contrast to an autotroph) but relies on organic carbon for growth and as an energy source.

Hibernation: A state of inactivity in animals characterized by reduced body temperature, breathing, and metabolic rates; usually occurs in response to cold winter conditions when food supplies are limited.

Homeotherm: An organism that maintains a stable internal body temperature, irrespective of external influences. The opposite is a poikilotherm, whose internal temperature can vary widely.

Humus: Organic matter that has cannot be further broken down under the same environmental conditions.

Hydrochory: The dispersal of plant seeds by water.

Hyphae: Long, branching filaments that make up the vegetative structure of most fungi. Together they make up the mycelium.

Ideal free distribution: A theory of how organisms distribute themselves among several resource-containing habitats. It states that individuals that are free to colonize any habitat and are globally aware of the pattern of resource distribution across all habitats will aggregate in each at a density that matches the proportional amount of resources available. Once the ideal distribution is reached, all organisms, no matter the original resource richness of their habitat or the current density of their competitors, will enjoy the same fitness.

Interference competition: A contest between organisms leading to a decline in the rate of resource exploitation that follows from direct interactions, often aggressive, among competitors.

Intraguild predation: Killing and eating potential competitors. This interaction includes both predation and competition because all the predator species use the same prey resources but can also feed on each other.

Landscape ecology: A branch of ecology that emphasizes the relation between patterns, processes, and scales, focusing on broad-scale ecological and environmental issues. Studies often consider large spatial scales and examine the relation between human or natural development and ecological processes.

Molting: A periodic casting off of part of an organism's body (often the outer layer) either at specific times of year or at particular points in its life cycle.

Monocots (monocotyledons): The group of flowering plants whose seeds have only one embryonic leaf (cotyledon).

Monogamous: Bearing the offspring of a single reproductive mate at any one time (genetic monogamy), or mating and having young with a single sexual partner (social monogamy).

Mutualism: An interaction between two species in which both benefit through greater survival and reproductive output (fitness).

Mycorrhizae: Usually mutualistic associations between a fungus and the roots of a vascular plant.

Myrmecochory: Plant seed dispersal by ants.

Natural selection: One of the key mechanisms of evolutionary theory whereby biological traits become either more or less common in a population according to the differential reproductive success of their bearers.

Nutrient: A chemical an organism needs to live and grow that must be taken in from the environment.

Object permanence: Understanding that objects continue to exist even when a barrier prevents them from being seen, heard, or touched.

Omnivore: A species that consumes prey from a variety of trophic levels, often one that eats both plants and animals.

Opportunist: A species that survives by rapidly colonizing habitats, usually because of high dispersal and reproductive capacities.

Organic: Coming from a once-living organism and capable of decaying (or the product of decay), or composed of carbon in biologically active molecules.

Oxidation: The chemical interaction between oxygen molecules and substances they react with, resulting in the loss of at least one electron. The opposite reaction is called reduction.

Palpi (pedipalpi or palps): A pair of appendages on spiders that can be used to identify species and that mature males use to transfer sperm to females.

Parasitism: A type of nonmutual relationship between species in which one organism (the parasite) benefits at the expense of the other (the host) without killing the host. Parasites are generally much smaller than their hosts.

Parasitoids: Similar to parasites (parasitism) except that parasitoids generally lay their eggs in their hosts' eggs or larvae, ultimately killing or sterilizing them.

Phenotypic plasticity: An organism's ability to change its phenotype (characteristics or traits) in response to changes in its environment. This generally excludes behavior as a phenotype (behavioral plasticity).

Pheromones: Chemicals that are secreted or excreted by an organism and trigger a response by members of the same species.

Photosynthesis: The chemical process carried out by plants, algae, and some bacteria that converts carbon dioxide into sugars using water and the energy from light and releasing oxygen as waste.

Physiological ecology: The branch of ecology that considers how organisms are physiologically adapted to their environment (mechanical, physical, and biochemical functions) and their responses to changes in it.

Phytoplankton: The microscopic autotrophic organisms that inhabit open-water regions of the oceans and freshwater bodies (a type of microalgae).

Plankton: Mostly small, relatively nonmotile organisms that inhabit the open regions of all water bodies, having plant components (phytoplankton) and animal components (zooplankton).

Plasticity. See *Phenotypic plasticity.*

Polymorphism: The stable or regular occurrence of two or more morphological forms of organisms within the same population of a species.

Population dynamics: The short- and long-term changes in the size and composition of single-species populations affected by birth, death, immigration, and emigration.

Population ecology: The field of ecology that specializes in the study of population dynamics.

Predation pressure: The effect of predation on a prey population, resulting in the decrease of that population and possibly the evolution of antipredator adaptations.

Primary consumers: Organisms that obtain energy and nutrients from primary producers (autotrophs). Also called herbivores.

Primary producers. See *Autotrophs.*

Sapwood: The newest, outermost wood in a tree that is living, which conducts water from the roots to the leaves. Also called alburnum.

Smog: Visible air pollution; a smokelike fog.

Social insect: A species of insect that lives in colonies with a division of labor, overlapping generations, and cooperative brood care.

Species richness: The total number of different species in a particular habitat or community.

Stomate: A pore in the epidermis of leaves and stems that is used for gas exchange and to release excess water vapor (transpiration).

Symbiosis: Most commonly, a persistent (long-term) mutualism.

Thallus: Undifferentiated vegetative tissue found in various groups of organisms, including fungi, algae, lichens, and liverworts.

Thermal energy: The total internal energy of a sample of matter that determines its temperature.

Thermoregulation: An organism's ability to keep its body temperature within certain boundaries, despite external temperatures. See also *Homeotherm*.

Torpor: A short-term state of decreased physiological activity in an animal, as in hibernation. Also called temporary hibernation.

Trophic level: The position an organism occupies in a food chain. Examples include autotroph, primary consumer, and secondary consumer.

Zoochory: The dispersion of plant seeds by animals.

Zooplankton: Animal plankton, generally consumers of phytoplankton and bacteria.